REVISED AND UPDATED

METALWORKER'S DATA BOOK

FOR HOME MACHINISTS

The Essential Reference Guide for
Everyone Who Works with Metal

HAROLD HALL

Revised and updated by **George Bulliss**,
editor of *The Home Shop Machinist* magazine

FOX CHAPEL
PUBLISHING

Parts of this book were updated for today's American reader with regard to new techniques and tools. These updates were graciously provided by George Bulliss of *The Home Shop Machinist*, *Machinist's Workshop*, and *Digital Machinist* magazines.

First published in the United Kingdom by Special Interest Model Books
© Special Interest Model Books Ltd 2009
First published in North America in 2017 by Fox Chapel Publishing, 903 Square Street, Mount Joy, PA 17552.

Cover image © Kadmy (Adobe Stock)

ISBN 978-1-56523-913-5

Library of Congress Cataloging-in-Publication Data

Names: Hall, Harold, 1933- author.
Title: Metalworker's data book for home machinists / Harold Hall.
Description: East Petersburg : Fox Chapel Publishing, 2017. | Includes
   bibliographical references and index.
Identifiers: LCCN 2017001917 | ISBN 9781565239135
Subjects: LCSH: Workshops--Handbooks, manuals, etc. | Metal-working
   machinery--Handbooks, manuals, etc.
Classification: LCC TT152 .H35 2017 | DDC 684/.08--dc23
LC record available at https://lccn.loc.gov/2017001917

To learn more about the other great books from Fox Chapel Publishing, or to find a retailer near you, call toll-free 800-457-9112 or visit us at *www.FoxChapelPublishing.com*.

Printed in Singapore
First printing

CONTENTS

PREFACE

The data required in the metalworking workshop covers a wide and varied range of subjects, a requirement that this data book seeks to fulfill. Of course, few readers will need all the data that is included, but no doubt all of it will be needed by someone. For my part, many of the subjects covered were new to me or, at least, were subjects with which I had only limited involvement in the past. Making this data available required in-depth study and I am therefore now much better informed on many subjects. This in itself made it a very gratifying task.

In these rapidly changing times, some of the content may seem superfluous because it relates to systems that are no longer widely used. However, this very fact makes the publication of these data even more important, because such information will become increasingly difficult to locate when the need for it arises, e.g. when restoring equipment made many years ago.

Harold Hall

All reasonable care has been taken in the preparation of this data book, but neither the publishers nor the author can be held legally responsible for errors in its content or for any loss, however arising, from such errors. Reliance placed upon these data is at the reader's own risk.

INTRODUCTION

The data in this book have not been extracted from the first available source because my experience has shown that errors exist in even the most reputable publications – and this book is unlikely to be an exception. The information has therefore been corroborated wherever possible by reference to more than one source. Unfortunately, information relating to earlier standards is not readily available in national standards, so reliance has had to be placed on data published elsewhere.

Mostly, the information given for a subject is sufficiently complete for normal workshop activities. However, in a very few cases, e.g. electric motors, the information given is only a 'lead-in' to the topic and more detailed information may be required.

Many of the values given are subject to tolerances, e.g. those for drill diameters, but the inclusion of these would have made the book overly complex and unwieldy. Nevertheless, where considered appropriate, some indication of tolerance requirements is given. On the rare occasion when a reader may need greater detail, he or she is advised to consult the international/national standards, or other appropriate literature.

The data in this book is intended to be what I would call 'hard and fast', i.e. data that is not subject to variations due to indefinable factors. Therefore, information such as machine tool-cutting speeds, feeds, tool-rake angles and the like, which depend on so many factors, typically machine rigidity, are not covered.

With the increasing use of calculators and computers, the conversion charts may seem superfluous. However, I believe that there are still some people who will find them useful, especially if they are working on a major project and need to convert a large number of dimensions.

Readers may wish to know when the data for this book were compiled, in order to consider their relevance at some future time. Much of the content was originally researched over the period 1994–2000 but, where necessary, it was checked against the latest standards in 2007/ 8. Most data will never change but, inevitably, some will become less relevant, e.g. details of Whitworth threads. However, as mentioned in the preface, such data will become increasingly difficult to locate, which more than justifies their inclusion in this book.

CHAPTER 1
DRILLS

The imperial drill ranges, fractional, number and letter, which were established many years ago, created a range of drills with irregular size increments. For example, if we take a letter K drill (0.2810in) and compare it with the next size up ⁹⁄₃₂in, (0.28125in) it is only ¼ of 1 thou. larger. However, the next size up from ⁹⁄₃₂ in is a letter L drill (0.2900in), which is almost 9 thou. larger. It is therefore suggested that readers should standardize on metric sizes in 0.1mm increments (approx. 0.004in).

STANDARD DRILL SIZES

Table 1.1 lists both imperial and metric standard sizes so that the size ranges can be easily compared. This will make it simpler to choose an alternative drill if the size required is not available in the workshop.

Table 1.1. Standard drills: metric and imperial sizes

Drill sizes			Decimal sizes	
Metric	Imperial			
mm	Fractions	No./Letter	mm	in
0.30	-	-	0.300	0.0118
0.32	-	-	0.320	0.0126
-	-	80	0.343	0.0135
0.35	-	-	0.350	0.0138
-	-	79	0.368	0.0145
0.38	-	-	0.380	0.0150
-	1/64	-	0.397	0.0156
0.40	-	-	0.400	0.0157
-	-	78	0.406	0.0160
0.42	-	-	0.420	0.0165
0.45	-	-	0.450	0.0177
-	-	77	0.457	0.0180
0.48	-	-	0.480	0.0189
0.50	-	-	0.500	0.0197
-	-	76	0.508	0.0200
0.52	-	-	0.520	0.0205

Drill sizes			Decimal sizes	
Metric	Imperial			
mm	Fractions	No./Letter	mm	in
-	-	75	0.533	0.0210
0.55	-	-	0.550	0.0217
-	-	74	0.571	0.0225
0.58	-	-	0.580	0.0228
0.60	-	-	0.600	0.0236
-	-	73	0.610	0.0240
0.62	-	-	0.620	0.0244
-	-	72	0.635	0.0250
0.65	-	-	0.650	0.0256
-	-	71	0.660	0.0260
0.68	-	-	0.680	0.0268
0.70	-	-	0.700	0.0276
-	-	70	0.711	0.0280
0.72	-	-	0.720	0.0283
-	-	69	0.742	0.0292
0.75	-	-	0.750	0.0295

Drill sizes				Decimal sizes	
Metric	Imperial				
mm	Fractions	No./Letter		mm	in
0.78	-	-		0.780	0.0307
-	-	68		0.787	0.0310
-	1/32	-		0.794	0.0313
0.80	-	-		0.800	0.0315
-	-	67		0.813	0.0320
0.82	-	-		0.820	0.0323
-	-	66		0.838	0.0330
0.85	-	-		0.850	0.0335
0.88	-	-		0.880	0.0346
-	-	65		0.889	0.0350
0.90	-	-		0.900	0.0354
-	-	64		0.914	0.0360
0.92	-	-		0.920	0.0362
-	-	63		0.940	0.0370
0.95	-	-		0.950	0.0374
-	-	62		0.965	0.0380
0.98	-	-		0.980	0.0386
-	-	61		0.991	0.0390
1.00	-	-		1.000	0.0394
-	-	60		1.016	0.0400
-	-	59		1.041	0.0410
1.05	-	-		1.050	0.0413
-	-	58		1.067	0.0420
-	-	57		1.092	0.0430
1.10	-	-		1.100	0.0433
1.15	-	-		1.150	0.0453
-	-	56		1.181	0.0465
-	3/64	-		1.191	0.0469
1.20	-	-		1.200	0.0472
1.25	-	-		1.250	0.0492
1.30	-	-		1.300	0.0512

Drill sizes				Decimal sizes	
Metric	Imperial				
mm	Fractions	No./Letter		mm	in
-	-	55		1.321	0.0520
1.35	-	-		1.350	0.0531
-	-	54		1.397	0.0550
1.40	-	-		1.400	0.0551
1.45	-	-		1.450	0.0571
1.50	-	-		1.500	0.0591
-	-	53		1.511	0.0595
1.55	-	-		1.550	0.0610
-	1/16	-		1.587	0.0625
1.60	-	-		1.600	0.0630
-	-	52		1.613	0.0635
1.65	-	-		1.650	0.0650
1.70	-	-		1.700	0.0669
-	-	51		1.702	0.0670
1.75	-	-		1.750	0.0689
-	-	50		1.778	0.0700
1.80	-	-		1.800	0.0709
1.85	-	-		1.850	0.0728
-	-	49		1.854	0.0730
1.90	-	-		1.900	0.0748
-	-	48		1.930	0.0760
1.95	-	-		1.950	0.0768
-	5/64	-		1.984	0.0781
-	-	47		1.994	0.0785
2.00	-	-		2.000	0.0787
2.05	-	-		2.050	0.0807
-	-	46		2.057	0.0810
-	-	45		2.083	0.0820
2.10	-	-		2.100	0.0827
2.15	-	-		2.150	0.0846
-	-	44		2.184	0.0860
2.20	-	-		2.200	0.0866

| Drill sizes | | | Decimal sizes | | | Drill sizes | | | Decimal sizes | |
| Metric | Imperial | | | | | Metric | Imperial | | | |
mm	Fractions	No./Letter	mm	in		mm	Fractions	No./Letter	mm	in
						-	-	31	3.048	0.1200
2.25	-	-	2.250	0.0886		3.10	-	-	3.100	0.1220
-	-	43	2.261	0.0890		-	1/8	-	3.175	0.1250
2.30	-	-	2.300	0.0906		3.20	-	-	3.200	0.1260
2.35	-	-	2.350	0.0925		3.25	-	-	3.250	0.1280
-	-	42	2.375	0.0935						
						-	-	30	3.264	0.1285
-	3/32	-	2.381	0.0938		3.30	-	-	3.300	0.1299
2.40	-	-	2.400	0.0945		3.40	-	-	3.400	0.1339
-	-	41	2.438	0.0960		-	-	29	3.454	0.1360
2.45	-	-	2.450	0.0965		3.50	-	-	3.500	0.1378
-	-	40	2.489	0.0980						
						-	-	28	3.569	0.1405
2.50	-	-	2.500	0.0984		-	9/64	-	3.572	0.1406
-	-	39	2.527	0.0995		3.60	-	-	3.600	0.1417
2.55	-	-	2.550	0.1004		-	-	27	3.658	0.1440
-	-	38	2.578	0.1015		3.70	-	-	3.700	0.1457
2.60	-	-	2.600	0.1024						
						-	-	26	3.734	0.1470
-	-	37	2.642	0.1040		3.75	-	-	3.750	0.1476
2.65	-	-	2.650	0.1043		-	-	25	3.797	0.1495
2.70	-	-	2.700	0.1063		3.80	-	-	3.800	0.1496
-	-	36	2.705	0.1065		-	-	24	3.861	0.1520
2.75	-	-	2.750	0.1083						
						3.90	-	-	3.900	0.1535
-	7/64	-	2.778	0.1094		-	-	23	3.912	0.1540
-	-	35	2.794	0.1100		-	5/32	-	3.969	0.1563
2.80	-	-	2.800	0.1102		-	-	22	3.988	0.1570
-	-	34	2.819	0.1110		4.00	-	-	4.000	0.1575
2.85	-	-	2.850	0.1122						
						-	-	21	4.039	0.1590
-	-	33	2.870	0.1130		-	-	20	4.089	0.1610
2.90	-	-	2.900	0.1142		4.10	-	-	4.100	0.1614
-	-	32	2.946	0.1160		4.20	-	-	4.200	0.1654
2.95	-	-	2.950	0.1161		-	-	19	4.216	0.1660
3.00	-	-	3.000	0.1181						
						4.25	-	-	4.250	0.1673

Drill sizes			Decimal sizes			Drill sizes			Decimal sizes	
Metric	Imperial					Metric	Imperial			
mm	Fractions	No./Letter	mm	in		mm	Fractions	No./Letter	mm	in
4.30	-	-	4.300	0.1693		5.50	-	-	5.500	0.2165
-	-	18	4.305	0.1695		-	7/32	-	5.556	0.2188
-	11/64	-	4.366	0.1719		5.60	-	-	5.600	0.2205
-	-	17	4.394	0.1730						
						-	-	2	5.613	0.2210
4.40	-	-	4.400	0.1732		5.70	-	-	5.700	0.2244
-	-	16	4.496	0.1770		5.75	-	-	5.750	0.2264
4.50	-	-	4.500	0.1772		-	-	1	5.791	0.2280
-	-	15	4.572	0.1800		5.80	-	-	5.800	0.2283
4.60	-	-	4.600	0.1811						
						5.90	-	-	5.900	0.2323
-	-	14	4.623	0.1820		-	-	A	5.944	0.2340
-	-	13	4.699	0.1850		-	15/64	-	5.953	0.2344
4.75	-	-	4.750	0.1870		6.00	-	-	6.000	0.2362
-	3/16	-	4.762	0.1875		-	-	B	6.045	0.2380
4.80	-	12	4.800	0.1890						
						6.10	-	-	6.100	0.2402
-	-	11	4.851	0.1910		-	-	C	6.147	0.2420
4.90	-	-	4.900	0.1929		6.20	-	-	6.200	0.2441
-	-	10	4.915	0.1935		-	-	D	6.248	0.2460
-	-	9	4.978	0.1960		6.25	-	-	6.250	0.2461
5.00	-	-	5.000	0.1969						
						6.30	-	-	6.300	0.2480
-	-	8	5.055	0.1990		-	1/4	E	6.350	0.2500
5.10	-	-	5.100	0.2008		6.40	-	-	6.400	0.2520
-	-	7	5.105	0.2010		6.50	-	-	6.500	0.2559
-	13/64	-	5.159	0.2031		-	-	F	6.528	0.2570
-	-	6	5.182	0.2040						
						6.60	-	-	6.600	0.2598
5.20	-	-	5.200	0.2047		-	-	G	6.629	0.2610
-	-	5	5.220	0.2055		6.70	-	-	6.700	0.2638
5.25	-	-	5.250	0.2067		-	17/64	-	6.747	0.2656
5.30	-	-	5.300	0.2087		6.75	-	-	6.750	0.2657
-	-	4	5.309	0.2090						
						-	-	H	6.756	0.2660
5.40	-	-	5.400	0.2126		6.80	-	-	6.800	0.2677
-	-	3	5.410	0.2130		6.90	-	-	6.900	0.2717

| Drill sizes | | | Decimal sizes | |
| Metric | Imperial | | | |
mm	Fractions	No./Letter	mm	in
-	-	I	6.909	0.2720
7.00	-	-	7.000	0.2756
-	-	J	7.036	0.2770
7.10	-	-	7.100	0.2795
-	-	K	7.137	0.2810
-	9/32	-	7.144	0.2813
7.20	-	-	7.200	0.2835
7.25	-	-	7.250	0.2854
7.30	-	-	7.300	0.2874
-	-	L	7.366	0.2900
7.40	-	-	7.400	0.2913
-	-	M	7.493	0.2950
7.50	-	-	7.500	0.2953
-	19/64	-	7.541	0.2969
7.60	-	-	7.600	0.2992
-	-	N	7.671	0.3020
7.70	-	-	7.700	0.3031
7.75	-	-	7.750	0.3051
7.80	-	-	7.800	0.3071
7.90	-	-	7.900	0.3110
-	5/16	-	7.938	0.3125
8.00	-	-	8.000	0.3150
-	-	O	8.026	0.3160
8.10	-	-	8.100	0.3189
8.20	-	-	8.200	0.3228
-	-	P	8.204	0.3230
8.25	-	-	8.250	0.3248
8.30	-	-	8.300	0.3268
-	21/64	-	8.334	0.3281
8.40	-	-	8.400	0.3307

| Drill sizes | | | Decimal sizes | |
| Metric | Imperial | | | |
mm	Fractions	No./Letter	mm	in
-	-	Q	8.433	0.3320
8.50	-	-	8.500	0.3346
8.60	-	-	8.600	0.3386
-	-	R	8.611	0.3390
8.70	-	-	8.700	0.3425
-	11/32	-	8.731	0.3438
8.75	-	-	8.750	0.3445
8.80	-	-	8.800	0.3465
-	-	S	8.839	0.3480
8.90	-	-	8.900	0.3504
9.00	-	-	9.000	0.3543
-	-	T	9.093	0.3580
9.10	-	-	9.100	0.3583
-	23/64	-	9.128	0.3594
9.20	-	-	9.200	0.3622
9.25	-	-	9.250	0.3642
9.30	-	-	9.300	0.3661
-	-	U	9.347	0.3680
9.40	-	-	9.400	0.3701
9.50	-	-	9.500	0.3740
-	3/8	-	9.525	0.3750
-	-	V	9.576	0.3770
9.60	-	-	9.600	0.3780
9.70	-	-	9.700	0.3819
9.75	-	-	9.750	0.3839
9.80	-	-	9.800	0.3858
-	-	W	9.804	0.3860
9.90	-	-	9.900	0.3898
-	25/64	-	9.922	0.3906
10.00	-	-	10.000	0.3937
-	-	X	10.084	0.3970

Drill sizes — Metric mm	Drill sizes — Imperial Fractions	Drill sizes — Imperial No./Letter	Decimal sizes mm	Decimal sizes in
10.10	-	-	10.100	0.3976
10.20	-	-	10.200	0.4016
10.25	-	-	10.250	0.4035
-	-	Y	10.262	0.4040
10.30	-	-	10.300	0.4055
-	13/32	-	10.319	0.4063
10.40	-	-	10.400	0.4094
-	-	Z	10.490	0.4130
10.50	-	-	10.500	0.4134
10.60	-	-	10.600	0.4173
10.70	-	-	10.700	0.4213
-	27/64	-	10.716	0.4219
10.75	-	-	10.750	0.4232
10.80	-	-	10.800	0.4252
10.90	-	-	10.900	0.4291
11.00	-	-	11.000	0.4331
11.10	-	-	11.100	0.4370
-	7/16	-	11.112	0.4375
11.20	-	-	11.200	0.4409
11.25	-	-	11.250	0.4429
11.30	-	-	11.300	0.4449
11.40	-	-	11.400	0.4488
11.50	-	-	11.500	0.4528
-	29/64	-	11.509	0.4531
11.60	-	-	11.600	0.4567
11.70	-	-	11.700	0.4606
11.75	-	-	11.750	0.4626
11.80	-	-	11.800	0.4646
11.90	-	-	11.900	0.4685
-	15/32	-	11.906	0.4688
12.00	-	-	12.000	0.4724

Drill sizes — Metric mm	Drill sizes — Imperial Fractions	Drill sizes — Imperial No./Letter	Decimal sizes mm	Decimal sizes in
12.10	-	-	12.100	0.4764
12.20	-	-	12.200	0.4803
12.25	-	-	12.250	0.4823
12.30	-	-	12.300	0.4843
-	31/64	-	12.303	0.4844
12.40	-	-	12.400	0.4882
12.50	-	-	12.500	0.4921
12.60	-	-	12.600	0.4961
12.70	1/2	-	12.700	0.5000
12.75	-	-	12.750	0.5020
12.80	-	-	12.800	0.5039
12.90	-	-	12.900	0.5079
13.00	-	-	13.000	0.5118
-	33/64	-	13.097	0.5156
13.10	-	-	13.100	0.5157
13.20	-	-	13.200	0.5197
13.25	-	-	13.250	0.5217
13.30	-	-	13.300	0.5236
13.40	-	-	13.400	0.5276
-	17/32	-	13.494	0.5313
13.50	-	-	13.500	0.5315
13.60	-	-	13.600	0.5354
13.70	-	-	13.700	0.5394
13.75	-	-	13.750	0.5413
13.80	-	-	13.800	0.5433
-	35/64	-	13.891	0.5469
13.90	-	-	13.900	0.5472
14.00	-	-	14.000	0.5512
14.25	-	-	14.250	0.5610
-	9/16	-	14.287	0.5625

Drill sizes			Decimal sizes	
Metric	Imperial			
mm	Fractions	No./Letter	mm	in
14.50	-	-	14.500	0.5709
-	37/64	-	14.684	0.5781
14.75	-	-	14.750	0.5807
15.00	-	-	15.000	0.5906
-	19/32	-	15.081	0.5938
15.25	-	-	15.250	0.6004
-	39/64	-	15.478	0.6094
15.50	-	-	15.500	0.6102
15.75	-	-	15.750	0.6201
-	5/8	-	15.875	0.6250
16.00	-	-	16.000	0.6299

CENTER DRILLS

COMMON FORM

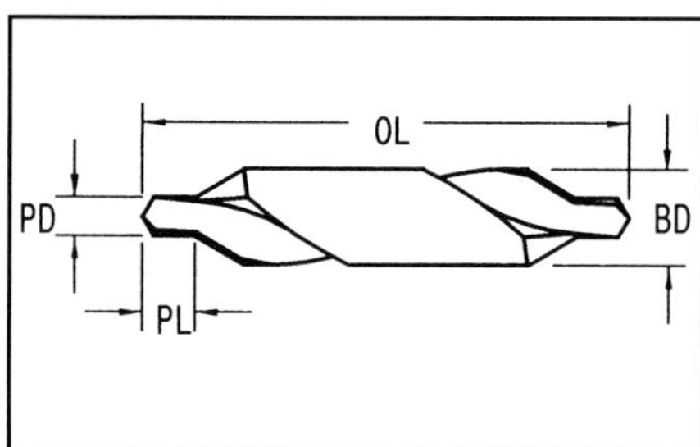

Point, included angle 118°
Countersink, included angle 60°

- BD — body diameter
- OL — overall length
- PD — point diameter
- PL — point length

Notes re tables

1. The lengths given are minimum values.

2. The body diameters of the bell-form center drill are larger than those listed.

3. Metric size center drills do not have a standard size reference and may be referred to by their body diameter (BD) or their point diameter (PD).

Sizes

Table 1.2 Metric center drill sizes (mm)

BD	OL	PD	PL
3.15	19.0	0.80	1.1
3.15	29.5	1.00	1.3
3.15	29.5	1.25	1.6
4.0	33.5	1.60	2.0
5.0	38.0	2.00	2.5
6.3	43.0	2.50	3.1
8.0	48.0	3.15	3.9
10.0	53.0	4.00	5.0
12.5	60.0	5.00	6.3
16.0	68.0	6.30	8.0
20.0	77.0	8.00	10.1

Above lengths may vary

Table 1.3 Imperial center drill sizes (in)

Size	BD	OL	PD	PL
BS1	1/8	1 1/2	3/64	1/16
BS2	3/16	1 3/4	1/16	5/64
BS3	1/4	2	3/32	1/8
BS4	5/16	2 1/4	1/8	5/32
BS5	7/16	2 1/2	3/16	1/4
BS5A	1/2	2 3/4	7/32	9/32
BS6	5/8	3	1/4	5/16
BS7	3/4	3 1/2	5/16	13/32

CENTER DRILL SECTIONS

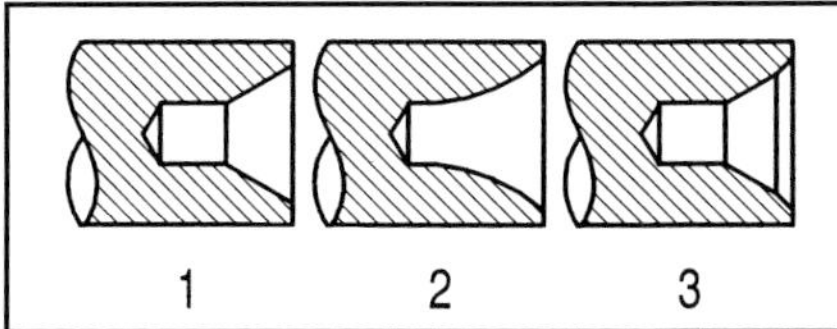

1. Center drill: this is the most common form.

2. Curved form: this makes only line contact with the lathe's center and is intended for light-duty precision work and where the tailstock center is set off-axis for taper-turning.

3. Bell form: this provides protection for the outer edge of the working countersink, which is important if the center is to be used for later operations, either in the manufacture of a part or in eventual refurbishment. This form is ideal for tooling used repeatedly between centers, e.g. between centers boring bars or work-holding mandrels.

Note: Forms 2 and 3 may only be available from specialist tool suppliers.

CHAPTER 2

TURNING TOOLS

REPLACEABLE TIP TOOLS

Replaceable tip tools conform to the standard dimensions set by the International Standards Organization (ISO) and, while minor variations exist between manufacturers in the case of the holders, inserts meet these standards exactly and are fully interchangeable.

INSERTS

The code for inserts consists of a 10-point numbering system and is used to define exactly the essential areas of dimension and therefore interchangeability. Code numbers between 1 and 7 must be given (but see '7 Radius' below, regarding round inserts). Code numbers 8–10 are used at the discretion of the manufacturer, who may add further codes for items such as the material from which the insert is made.

Inserts code:

1	Shape
2	Clearance angle
3	Tolerance
4	Type
5	Cutting edge length
6	Thickness
7	Radius
8–10	Used at manufacturer's discretion

1 SHAPE

Shape is coded by letter, as shown in the drawing.

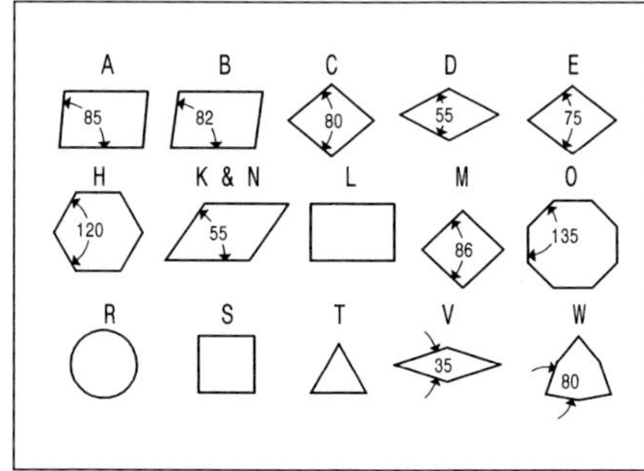

2 CLEARANCE ANGLE

The clearance angle is the angle for the insert itself. In practice, this maybe increased or decreased by angles included within the holder.

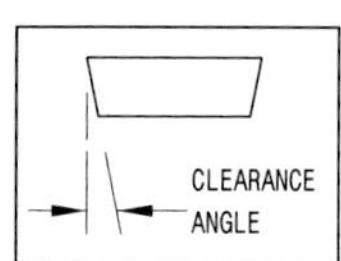

Clearance angle code:

A	3°
B	5°
C	7°
D	15°
E	20°
F	25°
G	30°
N	0°
P	11°
O	'Specials'

3 TOLERANCE

The tolerance classes are: A, C, E, F, G, H, J, K, L, M, N and U. Typically, these cover the accuracy of the tip's external dimensions and permit a tip to be replaced while still maintaining the dimensions of the parts being made. The details are not given here because they are complex and of little importance to the small workshop.

4 TYPE

The presence or absence of a central hole, with or without a countersink (CSK), and/or chip breakers are indicated by an alphabetical code (see drawing). X (not shown) is used for a 'Special.'

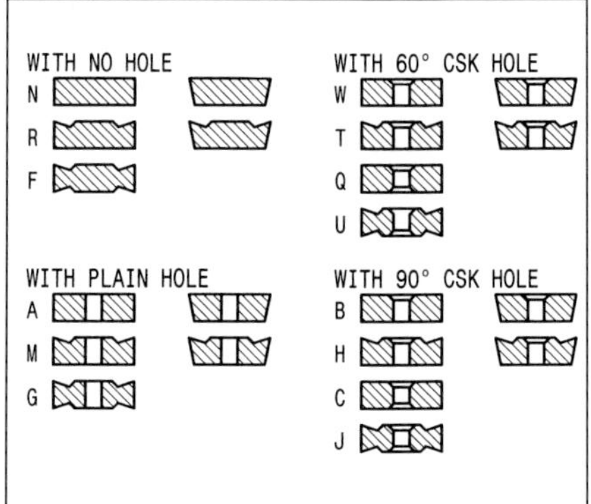

5 CUTTING EDGE LENGTH

This refers to the length of the cutting edge in millimeters, but ignoring any decimals, e.g. 12.5mm is given as 12. Single-digit lengths are prefixed with a zero, e.g. 9mm is given as 09. Inserts with different cutting edge lengths, typically a rectangle, are defined by the longer length. In the case of round inserts, the value is equal to its diameter.

6 THICKNESS

This is actually the height of the cutting edge above the tip's mounting surface and not the thickness of the main body of the tip that in some cases may differ.

Thickness code:

02	2.38mm
03	3.18mm
T3	3.97mm
04	4.76mm
05	5.56mm
06	6.35mm
07	7.94mm
09	9.52mm

7 RADIUS

This refers to the value of the radius between two adjacent cutting edges, but with the omission of the decimal point, e.g. 02 for 0.2mm and 16 for 1.6mm. In addition, 00 and M0 are used to indicate a round insert. However, as this feature is also defined by an 'R' in position 1 of the code, some manufacturers appear not to include this in position 7.

TOOL-HOLDERS: OUTSIDE DIAMETER TURNING TOOLS

Tool-holders are defined by a 9-point code, which must be given in every case. It has some features in common with the code for 'Inserts' (see above), but they do not occur in the same position within the code, e.g. 'Shape' is the first item in the 'Inserts code' but the second item in the 'Holder code.'

Some manufacturers also add a prefix of one or more letters/numbers. However, by looking at the whole code, it is usually easy to determine at which point the following code begins.

Holder code:

1	Insert locking system
2	Insert shape
3	Tool style
4	Insert clearance
5	Hand of tool
6	Shank height
7	Shank width
8	Tool length
9	Cutting edge length

1 INSERT LOCKING SYSTEM

Although interchangeability between makes of insert are guaranteed, the precise method of achieving locking methods differ between manufacturers. There are therefore differences in the holders produced by different manufacturers.

Insert locking system code:

S	Central locking screw; insert with countersunk hole
M	Central locating pin with side and top clamp; insert with parallel hole
P	Central locating pin and side clamp only; insert with parallel hole
C	Top clamp; insert without hole

The central screw (S) is likely to be the most common in the small workshop. With side clamping (M and P), the force may be achieved by a moving locating pin and a fixed opposing face or by a fixed locating pin and a moving opposing face.

2 INSERT SHAPE

Code: as for 'Inserts: 1 Shape' (see previous).

3 TOOL STYLE

The styles of holder are indicated by letters, as shown in the drawing. In each

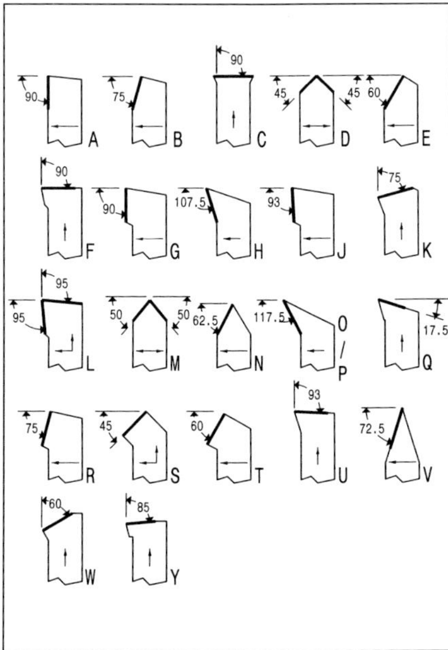

case, the bold line represents the main cutting edge(s) and the arrow(s) indicate the intended feed direction(s). However, feed in other directions is frequently possible, depending on the shape of the tip fitted. For example, the cutting face on a style 'R' holder can be achieved with either a square tip (S) or a triangular tip (T), or even others. The positions of secondary edges therefore vary but may be usable in some cases.

In addition, 'X' (not shown) can be used to indicate a boring tool used for back boring. This is where boring takes place with the tool being removed from the bore rather than entering it and will frequently be used to produce a step in the bore when in this case the tool is not fully withdrawn. 'X' may also be used for other specials.

4 INSERT CLEARANCE

Code: as for 'Inserts: 2 Clearance angle' (see previous).

5 HAND OF TOOL

Code:

R	Right-hand
L	Left-hand
N	Neutral; for left- or right-hand use.

6 SHANK HEIGHT

This is the height of the cutting edge above the base of the shank. It is frequently equal to the height of the shank itself, and is represented as, e.g.:

08	8mm
12	12mm.

7 SHANK WIDTH

The width of the shank is represented as, e.g.:

08	8mm
12	12mm.

8 TOOL LENGTH

This is the dimension from the tip of the insert to the end of the shank.

Tool length code:

A	32mm
B	40mm
C	50mm
D	60mm
E	70mm
F	80mm

G	90mm
H	100mm
J	110mm
K	125mm
L	140mm
M	150mm

P and above are longer.

X is used for a 'Special.'

9 CUTTING EDGE LENGTH

Code: as for 'Inserts, 5 Cutting edge length' (see previous).

BORING BARS

Boring bars are defined by a similar code to 'Outside diameter turning tools' but the entries are listed in a different order.

Boring bar code:

1	Shank type
2	Shank diameter
3	Tool length
4	Tip locking system
5	Insert shape
6	Tool style
7	Insert clearance
8	Hand of tool
9	Cutting edge length

1 SHANK TYPE

Code:

S	Solid steel
C	Solid carbide

2 SHANK DIAMETER

Code: this refers to the diameter of the shank, but it may also have flats to enable it to be mounted on to a flat surface. It is represented as, e.g:

08	8mm
12	12mm

3 TOOL LENGTH

Code: as for 'Outside diameter turning tools: 8 Tool length.'

4 TIP-LOCKING SYSTEM

Code: as for 'Outside diameter turning tools: 1 Insert locking system.'

5 INSERT SHAPE

Code: as for 'Inserts: 1 Shape.'

6 TOOL STYLE

Code: as for 'Inserts: 1 Shape.' See also 'Outside diameter turning tools: 3 Tool styles.'

7 INSERT CLEARANCE

Code as for 'Inserts: 2 Clearance angle.'

8 HAND OF TOOL

This code uses R and L as follows:

R   for conventional boring, where the workpiece is rotating counterclockwise

L   used where the workpiece is rotating clockwise.

9 CUTTING EDGE LENGTH

Code: as for 'Inserts: 5 Cutting edge length.'

A TYPICAL CODING

Insert	CCMT 060208
Tool-holder	SCLCR 1010E06
Boring bar	08K SCLCR06

Manufacturer's code e.g. UR H13A; this is added to the end of an ISO code and should be separated by a dash, e.g. CCMT 060208-UR H13A

IMPERIAL INSERTS

In some countries, inserts made to imperial dimensions are still assigned a code which is identical in principle to those defined above. However, the dimensions are given in the following ways:

Lengths are mostly given in multiples of $\frac{1}{8}$in, e.g.:

4	$\frac{1}{2}$in
6	$\frac{3}{4}$in

In a few cases they are given in multiples of $\frac{1}{4}$in.

Corner radius is given in multiples of $\frac{1}{64}$in, e.g.:

2	$\frac{1}{32}$in
3	$\frac{3}{64}$in

If in doubt, readers should consult the country's national standards..

BRAZED-TIP TUNGSTEN CARBIDE LATHE TOOLS

Brazed-tip tungsten carbide lathe tools are only available in a very limited range of shapes. They have a number of advantages compared with replaceable inserts, although these are fairly minor, e.g.:

- The initial cost is less.
- They can be made smaller, although, due to reductions in the size of replaceable tipped tooling, this is becoming less the case.
- They can be sharpened, although this makes them less convenient to use.

STANDARD SHAPES

A range of standard shapes is laid down by the BSI/ISO and DIN, some of which are marked with DIN numbers only (see list and drawing). There are also a few other common shapes that are not covered by these standards.

ISO1	DIN4971	Bar turning tool
ISO2	DIN4972	Cranked turning & facing tool
ISO3	DIN4978	Finishing tool
ISO4	DIN4976	Recessing tool
ISO5	DIN4977	Step turning and facing tool
ISO6	DIN4980	Cranked knife tool
ISO7	DIN4981	Parting off tool
ISO8	DIN4973	Rough and through boring tool
ISO9	DIN4974	Blind hole boring tool
ISO10	DIN4975	Chamfer tool
-	DIN282	External threading tool, 60>
-	DIN283	Internal threading tool, 60>

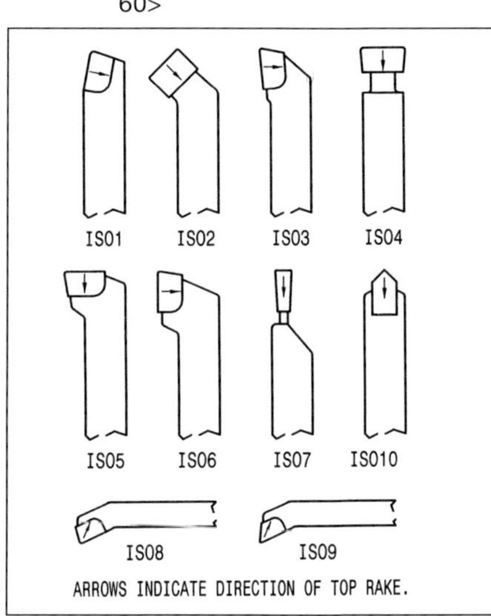

Common ISO shapes

02-05: Common ISO shapes>

Other common shapes

These include:

- Round-nosed tools
- Cranked round-nosed tools.

HAND OF TOOLS

For appropriate shapes, such as ISO6, both left- and right-handed tools are made.

Shank sizes

Except for parting-off tools that have rectangular shanks, shanks are square in section. The sizes are 8mm, 10mm, 12mm, 16mm and larger.

TUNGSTEN CARBIDE GRADES

The grades of tungsten carbide used for tipped tools are defined in a BSI/ISO standard. This lays down the capability of the grades, not their composition. It is necessary to understand the grading because it is frequently referred to in suppliers' catalogs.

CODE

The code for these grades consists of one letter and a two-digit number, e.g. P30. There are three letters (P, K and M) and six numbers (01, 10, 20, 30, 40 and 50), although not all of the numbers are used with each letter. A color code may also be used.

Letters

The letters indicate the type of material intended to be machined with the tool, i.e.:

P	Ferrous metals with long chips, such as steel, steel castings, malleable cast iron, etc.
K	Ferrous metals with short chips, such as gray cast iron and

hardened steel, also non-ferrous metals and non-metallic substances.

M   Ferrous metals with long or short chips and non-ferrous metals.

Numbers

The numbers indicate a progression in the degree of hardness and toughness of a tool. As the numbers increase, hardness decreases but toughness increases. They indicate the use to which the tool is best suited, e.g. finishing cuts, roughing cuts.

Color code

Tools may also be color-coded according to material, i.e.:

Blue   P
Red   K
Yellow M

Standard grades

The standard grades listed are:
P01, P10, P20, P30, P40 and P50
K01, K10, K20, K30 and K40
M10, M20, M30 and M40.

USES

The lower-number grades, being harder, are ideal for finishing cuts at higher speeds The higher number grades, being tougher, are better suited for roughing cuts at slower speeds and greater feed rates, and for difficult materials and conditions, e.g. intermittent cuts, castings with irregular and hard skins, use on a shaper

GENERAL-PURPOSE GRADES

In the small workshop, a tool with a middle-of-the-road value (i.e. P20 or P30) will suffice for general purposes. Consequently, most suppliers list only P20 or P30, although some also list K20 for use with cast-iron and non-ferrous materials. This is a perfectly adequate compromise for most uses in the average workshop.

CHAPTER 3
END MILLS AND SLOT DRILLS

SCREWED SHANKS

Screwed shank end mills and slot drills are, except for mini mills, by far the most common form of end mill and slot drill in the UK (though sometimes found, this form is somewhat rare in North America). They are made in a range of metric sizes, from 2.5mm to 50mm, with a similar range of imperial sizes. There are six shank sizes, each covering a range of cutter diameters. In some cases, the cutter diameter is larger than the shank diameter.

It is important to be aware of the cutter diameter/shank diameter combinations when purchasing cutters to ensure that they match the chuck collets available. Sizes up to 16mm shank diameter, which are those of most interest to the small workshop, are listed below.

SIZES: METRIC

Shank diameters are 6mm, 10mm, 12mm, 16mm and above, while cutters are available in increments of 0.5mm at the smaller diameters and of 1mm above (see Table 3.1).

Table 3.1 Metric shank and cutter sizes

Shank diameter (mm)	Cutter size (mm)
6	1.5–6
10	6.5–10
12	10.5–14
16	15–20

Longer end mills and slot drills are made and, apart from the omission of a few sizes, particularly the smaller sizes, the range follows that for standard length cutters.

SIZES: IMPERIAL

Imperial shank diameters are ¼in, ⅜in, ½in and ⅝in and above, while cutters are available in increments of ¹⁄₃₂in (see Table 3.2).

Table 3.2 Imperial shank and cutter sizes

Shank diameter (in)	Cutter size (in)
1/4	1/16–1/4
3/8	9/32–13/32
1/2	7/16–9/16
5/8	5/8–3/4

AVAILABILITY

Not all suppliers stock the full range of sizes; typically only cutters in 1mm or ¹⁄₁₆in increments are stocked.

THREAD SIZES

Both metric and imperial cutters have the same 20 TPI Whitworth form imperial thread. The diameter of the thread equals that of the shank diameter in both imperial and metric.

PLAIN SHANKS

End mills and slot drills are also made with plain shanks, sometimes with a flat for use with a screw to give added drive. While in the minority at sizes above 6mm diameter, the greater availability of suitable chucks for holding them now make them a more acceptable alternative. However, screwed shank mills are still to be preferred in most cases.

MINI MILLS

Although very common, the 6mm and ¼in plain shank mills (with flat) are not intended to be re-sharpened. They are therefore often referred to as 'disposable cutters', or sometimes as 'throw away' and 'mini mills'.

Unlike end mills (4-flute) and slot drills (2-flute), these mills have 3 flutes. One cutting tooth cuts over the center, making it suitable for plunge cutting. They are therefore equally suitable for use as end mills or slot drills.

SIZES: METRIC

Available sizes range from 1mm to 6mm, in 0.5mm increments. All have a shank diameter of 6mm and a flat for use with a drive screw. Other sizes, e.g. 2.8mm and 3.8mm, are only available from specialist suppliers, as are mills with a shank diameter of 10mm.

SIZES: IMPERIAL

Available sizes available range from ¹⁄₁₆in to ¼in, in ¹⁄₃₂in increments. All have a shank diameter of ¼in and a flat for use with a drive screw. Sizes in increments of ¹⁄₆₄in, are also available, again from specialist suppliers.

SCREWED SHANK DOVETAIL CUTTERS

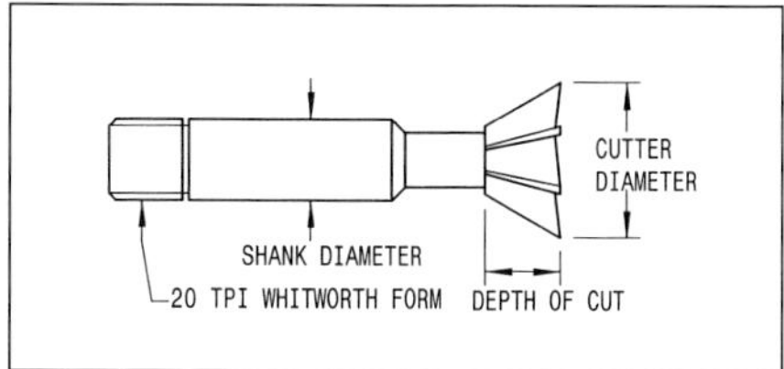

Dovetail cutters (see drawing) are made with screwed shanks of identical proportions and thread form to those on end mills, and also with plain shanks with a flat. The depth of cut increases with diameter, so it is necessary to obtain the correct diameter to suit the depth of dovetail to be machined.

CUTTER ANGLE

Cutters are commonly available with 60° and 45° angles, the 60° angle being by far the most common for machine slides. Cutters with a 55° angle are also listed but in a very limited range of sizes, and they are only available from specialist suppliers.

SIZES: IMPERIAL

Table 3.3 Depth of cut: 60° cutters

Shank diameter (in)	Cutter diameter (in)	Depth of cut (in)
1/2	1/2	5/32
1/2	5/8	7/32
1/2	3/4	9/32
1/2	7/8	3/8
1/2	1	15/32

Table 3.4 Depth of cut: 45° cutters

Shank diameter (in)	Cutter diameter (in)	Depth of cut (in)
1/2	1/2	1/8
1/2	5/8	5/32
1/2	3/4	7/32
1/2	7/8	1/4
1/2	1	5/16

SIZES: METRIC

Table 3.5 Depth of cut: 60° cutters

Shank diameter (mm)	Cutter diameter (mm)	Depth of cut (mm)
12	13	4
12	16	5.5
12	19	7
12	22	9.5
12	25	12

Table 3.6 Depth of cut: 45°cutters

Shank diameter (mm)	Cutter diameter (mm)	Depth of cut (mm)
12	13	3
12	16	4
12	19	5.5
12	22	6.5
12	25	7.5

CHAPTER 4

GRINDING WHEELS

MARKINGS

There is a standard method of marking grinding wheels which, in most cases, can be relied on to provide an accurate definition of a wheel's characteristics. It consists of a series of numbers and letters with well-defined meanings. Prefixes and/or suffixes may be added to provide additional data, as considered necessary by the manufacturer.

The meanings of the characters that make up the marking are described below, and are listed in the order in which they appear.

ABRASIVE

These fall into two categories:

A	aluminum oxide
C	silicon carbide.

Some manufacturers have their own trade names for these abrasives and use prefixes (e.g. BA, PA)to define small variations. For example, green silicon carbide, recommended for sharpening carbide-tipped tools, is usually, but not always, marked with the prefix 'G' (i.e. GC).

GRAIN SIZE

This is defined by a one-, two- or three-digit number; the smaller the number the larger the grain size. For a general-purpose off-hand grinder, a coarse wheel with a grain size of 36 (0.5mm average diameter) and a fine wheel with a grain size of 60 (0.25mm average diameter) would suit most applications.

GRADE

This is frequently defined as the hardness of the wheel and should not be confused with the hardness of the abrasive. It is the strength by which the individual grains are held together and therefore refers to the hardness of the wheel as a whole. It is defined by a letter on a scale of A to Z, with A being very soft and Z being very hard, although these extremes are rare. As an indication, H is considered soft, N medium and S hard.

STRUCTURE

This is best thought of as the porosity of the wheel because it defines the degree of open spaces between grains. It is represented by a number on a scale of 0 to 15, with 0 being dense and 15 being open. However, this characteristic is not always included in the marking.

BOND

The material which bonds the grains together is defined by this characteristic. There are five materials, each defined by a single letter:

V	vitrified
B	resinoid
R	rubber
E	shellac
S	silicate.

Vitrified is by far the most common. Resinoid, rubber and shellac are used to make thin wheels for cut-off machines, among other things. Silicate is for very special applications. Individual manufacturers may add additional characters to distinguish between small differences in bond method or material.

TYPICAL MARKING

The following is an example of a typical marking:

BA 46 N 8 V

ADDITIONAL MARKINGS

SPEED

With the exception of smaller wheels, all new wheels are also marked with the maximum permissible speed. Under no circumstances should this be exceeded. If a wheel is not marked, then do not use it unless you have other means of being absolutely sure of its maximum permissible speed.

BORE

The markings also include the bore diameter of the wheel. It is common for a wheel to have a bore larger than needed, requiring an adapter bush to be fitted. Do not use the wheel unless it is a close fit on the spindle on which it is to be used, with an adapter if necessary. Running a wheel with excessive clearance between bore and spindle is very dangerous. Do not do it!

CHAPTER 5

COLLETS AND TAPERS

COLLET CHUCKS

So many collet systems have been available in the past that it is only possible to detail in depth those that are currently relatively common.

TYPES

Collet chucks can be divided into a number of different groups, although there is considerable overlap between them. These are:

- Angle: very shallow; shallow; wide.
- Slotted: from one end; from both ends. (from page 27)
- Closing method: draw in; push in. (from page 28)
- Collet positioning: moving collet-fixed internal taper; fixed collet-moving internal taper.
- Individual collet gripping range: limited; wide.
- Purpose: work-holding; tool-holding.

ANGLE

Very shallow

These are mostly used in machine spindle tapers, typically a Morse taper. They are capable of a high gripping force, due to their shallow angle. The collet will hold itself in the taper, so therefore the closing ring, if used, has to be attached to the collet to draw it out.

These collets are mainly used in lathe spindles for workpiece-holding but their high gripping capability makes them ideal for holding milling cutters. As the collet

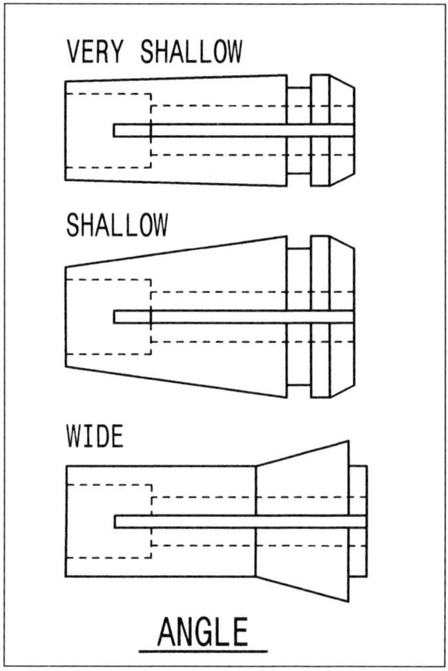

is almost entirely within the machine spindle, the overhang of the cutter is reduced compared to other forms of collet system. This produces a much more robust mounting, which is a definite plus when holding milling cutters.

As an alternative to a closing ring, a draw bar may be used. This is tapped gently on its end to release the collet from its taper. A major disadvantage of this type of collet is that it will not accept bar material.

Shallow

The shallow angle makes these collets ideally suited to tool-holding because of their gripping capability, but they can also be used for work-holding. Internal angles are typically around 16° and, in most cases, a closing ring is attached to the collet. By far the most common version is the ER range of collets(see below).

One rather specialized collet of this type, used in the past and still in use on some machines, is the Clare collet. This has the closing ring attached to the collet with a thread on the collet's outer diameter. The collet has a thread pitch slightly smaller than the closing thread that mates with the collet chuck body. The difference creates the effect of a much finer thread on the closing ring, so that less torque is required when tightening.

Also in this category, with an internal angle of 16° 50', are the collets that mate with R8 machine tapers and use the machine's draw bar rather than a closing ring.

Wide

These collets are mostly intended for work-holding because their wide angle gives them a relatively poor grip, making them unsuitable for the arduous task of end-mill holding. An internal angle of 20° is used on 5C collets, which are widely used for work-holding applications.

Note

The angles quoted are the total internal angles, e.g. 20° relates to 10° relative to the axis of the collet.

SLOTTED

From one end

This method is used on most collets.

From both ends

Shallow-angle collets are often slotted from either end, resulting in the bore remaining parallel when closing, giving a better grip

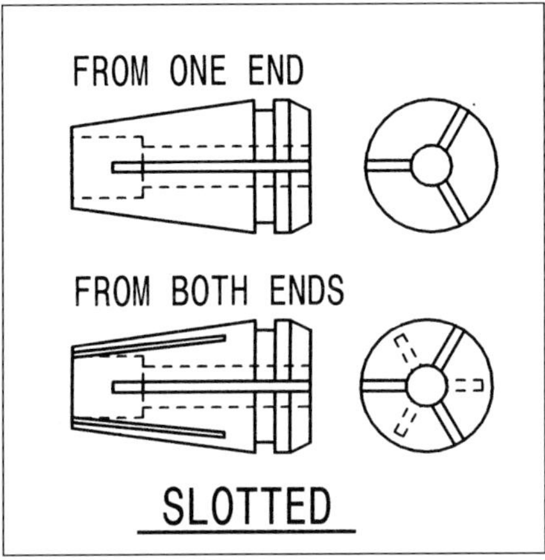

for tool-holding. Some are multi-slotted from either end, providing a wider gripping range from a single collet.

CLOSING METHOD

Draw in

This method is used on a few collet types where a tube or rod, threaded on to or into the rear of the collet, draws the collet into its mating taper.

Push in

This common method uses a closing ring to force the collet into its mating taper.

DEAD LENGTH

When a collet is being forced into a fixed taper, its final position will depend on the diameter of the workpiece. This makes secondary operation work that locates off the end of the collet rather inaccurate.

A collet system in which the outer ring is fixed and the internal taper moves forward to close the collet overcomes this problem. This system is not widely available and can be relatively expensive.

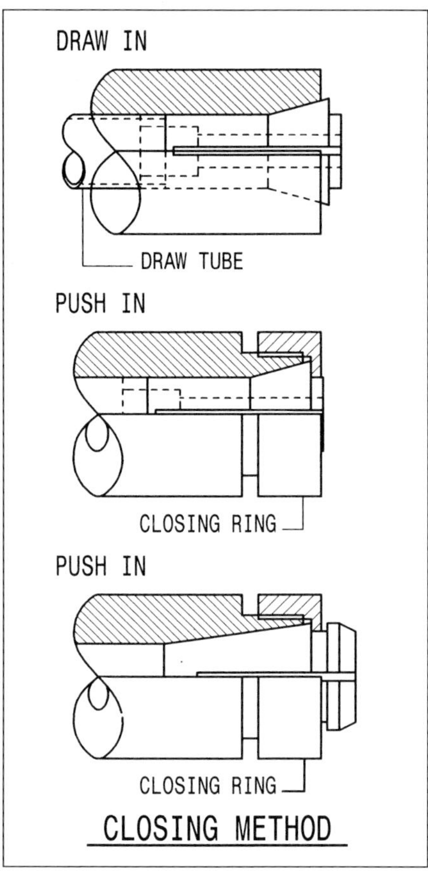

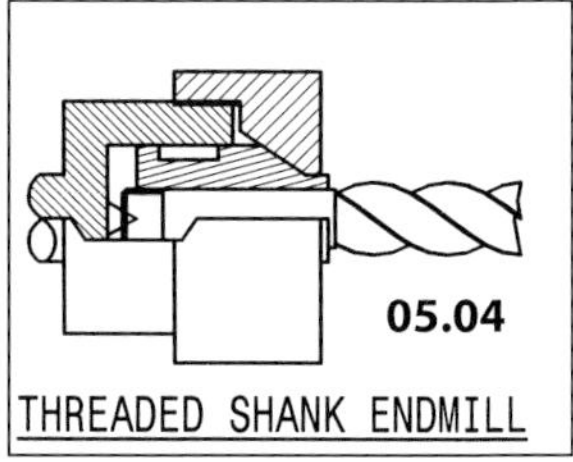

COLLETS FOR USE WITH THREADED SHANK END MILLS

The operation of collets for use with threaded shank end mills is best understood by reference to the drawing 05.04.

In operation, the closing ring is only lightly tightened and the end mill is then turned in the collet to bring the end of the end mill in contact with the base of the chuck bore. This frequently takes the form of a pointed center that locates in the center socket in the end of the end mill.

When a cutting load is placed on the end mill it may rotate very slightly in the collet thread, thus pulling the collet forward into the closing ring. However, the end mill will remain in the same axial position because it stays in contact with the base. This is a major advantage of this form of collet for end-mill holding, as the cutter's helix angle can result in it being pulled out of a plain collet.

COMMON TYPES

The individual collet types discussed below are those that were most common at the time this book was being written (2008). Unless stated otherwise, the collets work over a very limited range (typically, they work just acceptably over a range of, e.g. −0.2mm to +0.0mm). The maximum gripped sizes quoted below are typical, but individual manufacturers may work to slightly different ranges. There is a comparable range of imperial sizes (but see comment regarding ER collets below).

Morse taper collets

These have the following maximum diameters:

No. 1   8mm
No. 2   12mm
No. 3   18mm

R8 collets

These are used in the R8 machine bore common in many medium size milling machines and therefore have a common body size but with a gripping range of up to 20mm diameter.

ER collets

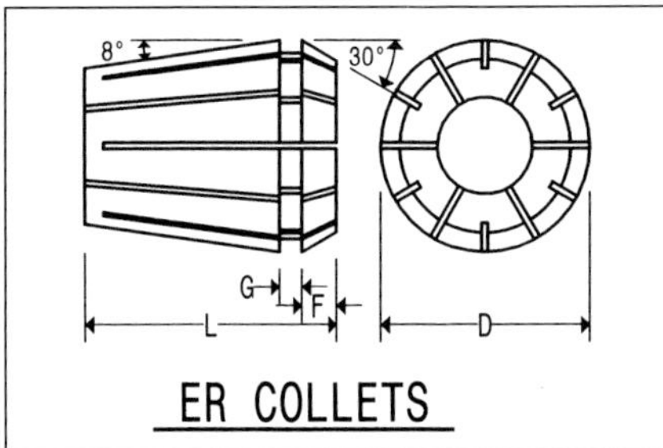

Table 5.1. ER collet dimensions

Size	Dimensions (mm)			
	D	L	G	F
ER11	11.5	18	2.0	2.5
ER16	17	27	2.7	4.0
ER20	21	31	2.8	4.8
ER25	26	35	3.1	5.0
ER32	33	40	3.6	5.5
ER40	41	46	4.1	7.0
ER50	52	60	5.5	8.5

The common sizes of ER collets are: ER11, ER16, ER20, ER25, ER32 and ER40 (a few manufacturers also list ER8 and ER50). For reference purposes, these collets are slightly larger than the maximum diameter (D); e.g. ER11 is 0.5mm larger, ER50 is 2mm larger, and the rest are 1mm larger. (These sizes probably refer to the maximum diameter of the taper in the chuck that holds the collets.)

Being multi-slotted from either end, these collets have a gripping range of −1.0mm to +0.0mm and are made in 1mm increments, although a few suppliers list the ER11 size in 0.5mm increments. Typically, a 10mm collet will hold workpieces between 9mm and 10mm in diameter, while an 11mm collet will grip workpieces between 10mm and 11mm. Therefore, given a complete set of collets, there are no gaps in the sizes of the workpieces that can be held, and the collets are also equally at home with imperial sizes despite their metric size reference.

The ER sizes of collets are only available with round bores and have the following maximum diameters:

ER11	7mm
ER16	10mm
ER20	13mm
ER25	16mm
ER32	20mm
ER40	26mm

5C collets

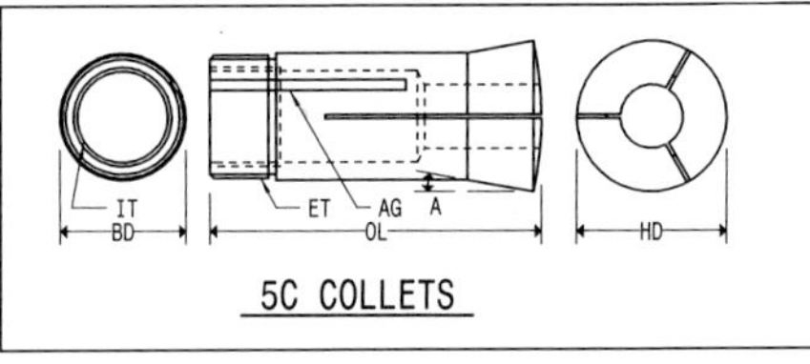

A	10°
AG	0.125 × 0.031 deep anti-turn groove
BD	1.250
ET	1.240 × 20 TPI external thread
HD	1 15/32
IT	1.040 × 24 TPI internal thread
OL	3 9/32

The 5C collet has a single body size with collets available up to a maximum gripping diameter of 30mm.

However, at the larger sizes, probably 25mm plus, the internal thread (IT) has to be sacrificed. Square and hexagonal collets are also available up to 18mm Sq/AF, or maybe a few millimeters more from some suppliers.

This collet can be used with an external closing ring but has both internal and external threads on the inner end, allowing it to be drawn in by various methods, including 'dead length' systems. It is used in a range of jigs and fixtures, including indexers, that are all widely available.

TAPERS

TYPES AND PURPOSE

The most common application for a taper in the home workshop is the machine spindle taper provided for tool-holding. By far the most common in this group are the Morse tapers, although the R8 taper is now becoming popular on milling machines.

Another form of taper is the one provided in the rear of drill chucks to enable them to be fitted to a taper shank (Morse, R8 etc.) or, in a few cases, direct to a drilling machine spindle. These tapers are known as Jacob's chuck tapers.

Taper pins are another taper found in the home workshop. These are useful for fitting items such as gears, pulleys and collars to spindles. (See page 123 for details of taper pins.)

MORSE TAPERS

Table 5.2. Morse tapers: dimensions

Taper no.	Gauge		Socket end (A)	Taper
	D	P		
0	0.252	2	0.356	0.62460
1	0.369	2 1/8	0.475	0.59858
2	0.572	2 9/16	0.700	0.59941
3	0.778	3 3/16	0.938	0.60235
4	1.020	4 1/16	1.231	0.62326
5	1.475	5 3/16	1.748	0.63151

Lengths/diameters in inches.
Tapers in inches per foot on diameter

Notes

1. The dimensions quoted are for the standard plug gauge.

2. The shank or socket may vary in length, in which case diameter D is maintained and the diameter at the larger end is allowed to change.

3. The base of the socket must be deep enough to accept shanks with a tang and must include provision for the extraction of the shank from the taper socket.

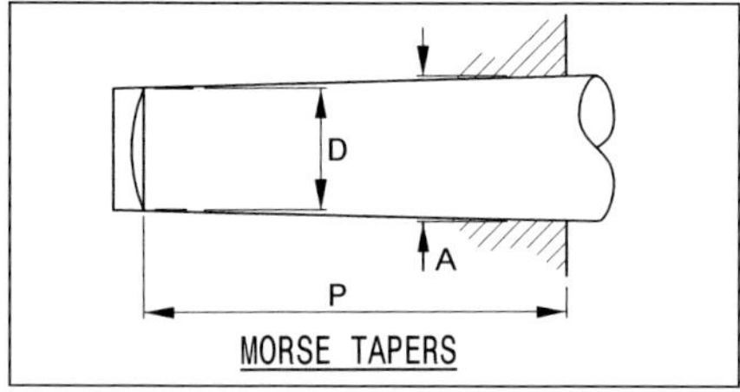

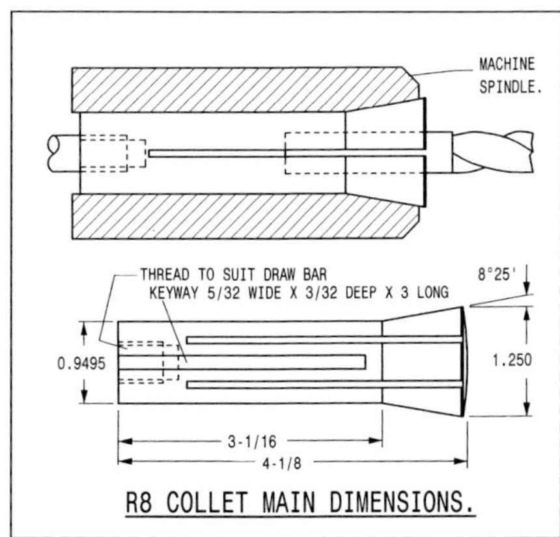

R8 TAPERS

The R8 taper is primarily used on milling machines, where its format permits the use of collets for holding end mills. This has the advantage of there being no overhang from the end of the machine spindle, as there would be with a conventional end-mill chuck.

The R8 taper can also be used with a solid shank, rather than a split collet, and it can be provided with an arbor, such as a Jacob's taper, for fitting a drill chuck.

The taper is not self-holding and must be used with a draw bar.

JACOB'S CHUCK TAPERS

Jacob's chuck tapers are internal tapers provided in precision and semi-precision drill chucks to enable them to be fitted to the required shank.

Economy drill chucks, as fitted to electric pistol drills, are provided with internal or external threads to facilitate their fitting. There are many thread sizes and the chuck being fitted should be checked for size.

Table 5.3. Jacob's chuck tapers: dimensions

No.	A	B	C	Taper
0	0.2500	0.22844	0.43750	0.59145
1	0.3840	0.33341	0.65625	0.92508
2	0.5590	0.48764	0.87500	0.97861
2a	0.5488	0.48764	0.75000	0.97861
3	0.8110	0.74610	1.21875	0.63898
4	1.1240	1.03720	1.65630	0.62886
5	1.4130	1.31610	1.87500	0.62010
6	0.6760	0.62410	1.00000	0.62292
33	0.6240	0.56050	1.00000	0.76194

Lengths/diameters in inches.
Taper in inches per foot on diameter.

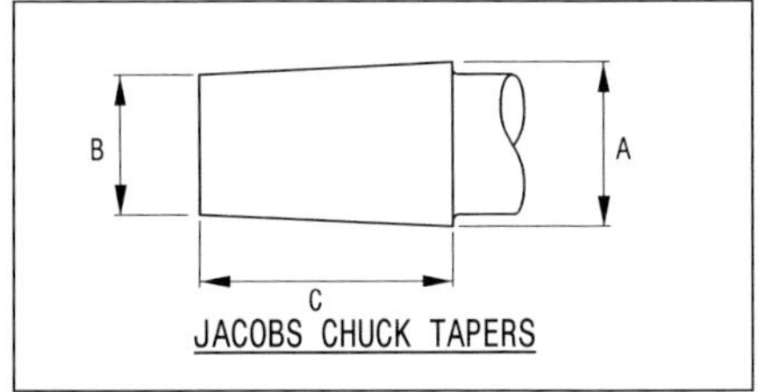

CHAPTER 6

PRECISION TOOLING

SLIP GAUGES

Slip gauges, also known as gauge blocks, or Johansson blocks, are metal blocks whose thicknesses are made to very precise dimensions. Supplied in sets of differing sizes, they can be stacked to give a wide range of overall dimensions.

SET SIZES

Sets consists of various numbers of blocks, although only one metric set and one imperial set are commonly available: an 88-piece metric set (occasionally 87, see comment below) and an 81-piece imperial set.

SET CONTENT

Each set is made up of a series of subsets, each providing a specific increment.

Metric 88-piece set

This consists of four subsets (making an 87-piece set) plus an optional additional block (making an 88-piece set):
- First subset: 9 blocks in increments of 0.001mm, ranging from 1.001mm to 1.009mm.
- Second subset: 49 blocks in increments of 0.01mm, ranging from 1.01mm to 1.49mm.
- Third subset: 19 blocks in increments of 0.5mm, ranging from 0.5mm to 9.5mm.
- Fourth subset: 10 blocks in increments of 10mm, ranging from 10mm to 100mm.

- Additional block: 1 block, of 1.0005mm, to provide an increment of 0.0005mm.

Imperial 81-piece set

This comprises four subsets:
- First subset: 9 blocks in increments of 0.0001in, ranging from 0.1001in to 0.1009in.
- Second subset: 49 blocks in increments of 0.001in, ranging from 0.101in to 0.149in.
- Third subset: 19 blocks in increments of 0.05in, ranging from 0.05in to 0.95in.
- Fourth subset: 4 blocks in increments of 1in, ranging from 1in to 4in.

ACCURACY

Slip gauges are made in various grades but only two are commonly available: Grade 1 and Grade 2 (in the case of imperial sets, these are known as 'Inspection' and 'Workshop'). Grade 2 is perfectly adequate for workshop use.

Accuracy is related to flatness, parallelism and length, with a permitted tolerance depending on the length of the gauge (longer gauges are allowed a slightly greater variation). A typical tolerance is within 0.00025mm.

MAKING UP GAUGE DIMENSIONS

In most cases, the dimension required can only be achieved by using a number of blocks rather than a single block. Ideally, the minimum number of blocks should be used. The following procedure, which satisfies the smallest increment at each stage, will establish a combination of blocks that give the required dimension. In most cases, the same result can be obtained with other combinations of blocks.

EXAMPLE REQUIRED DIMENSION: 23.123MM.

Procedure:

1. Satisfy the right-hand digit (of 23.123mm) with a single block, size 1.003mm, leaving 22.12mm.

2. Satisfy the right-hand digit (of 22.12mm) with a single block, size 1.02mm, leaving 21.1mm.

3. Satisfy the right-hand digit (of 21.1mm) with a single block, size 1.1mm, leaving 20mm.

4. The size of the fourth block will therefore be 20mm.

In most cases, more than one gauge will satisfy the right-hand digit. For example, in the above, 1.02 was chosen to satisfy the right-hand digit in 22.12mm, but 1.12, 1.22, etc. could also have been used. If the chosen block leaves a size that is difficult to achieve, alternative blocks should be considered.

DUPLICATE DIMENSIONS

Occasionally two sets of gauges will be required at the same time for the same dimension. In this case, a single block will no longer satisfy the right-hand digit so two blocks need to be used. In the example above, the right-hand digit of the initial dimension (23.123) can be satisfied by a combination of 1.006mm and 1.007mm blocks (which give a total of 2.013mm), leaving 21.11mm to be satisfied by the above process.

Sets with a smaller number of blocks, e.g. 47 metric and 41 imperial, can be used to achieve the same range of sizes, but they may be limited when it comes to producing duplicate dimensions.

SINE BARS

Sine bars can be used for setting up angles to a level of accuracy far better than other methods. The drawing shows how the angle (A) is dependent on the length of the sine bar (L) and the height of the packing (H). The two cylinders are part of the sine bar and are permanently attached to it. However, accuracy deteriorates as the angle increases, so this is not a practical method for angles much over 60°, although this will depend on the level of accuracy required.

Sine bar lengths are normally 125mm, 250mm, 5in or 10in, the dimension being the distance between the centers of the two built-in cylinders.

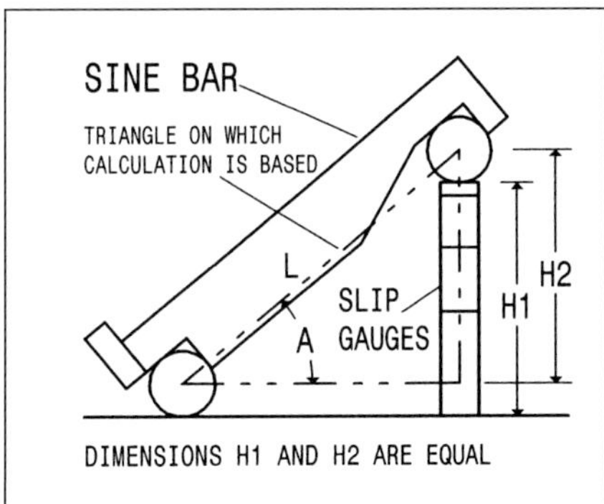

$H1 = \text{sine } A \times L$

where:

H = height of packing

A = angle

L = sine bar length (not labeled)

CHAPTER 7

SPANNERS

SIZES

Table 7.1 below lists spanners in increasing order of size.

NOTES

1. The sizes of Whitworth heads were reduced in around 1940. As a result, the head sizes on earlier equipment and older spanners may be one size larger.

2. In the Unified system, some sizes of bolt head (B) and nut (N) differ for a given thread size, the nut being the larger.

3. In the Unified system, some sizes are also made in a heavy range that has larger heads, usually one size larger.

4. The AF (across flats) figures quoted in the table are the nominal sizes for the head of the screw or the nut. The spanner itself will be slightly larger.

5. Using a spanner from one range on a hexagon from another range is not recommended. However, as the table shows, in a very few cases the metric and imperial sizes are very similar, e.g. a ¾in AF spanner is only 0.05mm (0.002in) larger than a 19mm spanner. In this case, it would probably be an acceptable alternative for a light-duty application, especially if a ring spanner or socket was being used. However, this approach should be used with considerable caution!

Table 7.1 Spanner sizes

Spanner size (AF)							
Decimal	Fractions	Metric					
in	in	mm		BA	Whit. /BSF	Unified	Metric
0.152	-	3.86		8	-	-	-
0.156	5/32	3.97		-	-	0 & 1	-
0.157	-	4.00		-	-	-	M2
0.172	-	4.37		7	-	-	-
0.187	3/16	4.76		-	-	2 & 3	-
0.193	-	4.90		6	-	-	-
0.197	-	5.00		-	-	-	M2.5
0.217	-	5.50		-	-	-	M3
0.220	-	5.59		5	-	-	-
0.236	-	6.00		-	-	-	-

Spanner size (AF)				Thread			
Decimal	Fractions	Metric					
in	in	mm		BA	Whit. /BSF	Unified	Metric
0.248	-	6.30		4	-	-	-
0.250	1/4	6.35		-	-	4	-
0.276	-	7.00		-	-	-	M4
0.282	-	7.16		3	-	-	-
0.312	5/16	7.94		-	-	5 & 6	-
0.315	-	8.00		-	-	-	M5
0.324	-	8.23		2	-	-	-
0.340	-	8.64		-	3/16	-	-
0.344	11/32	8.73		-	-	8	-
0.354	-	9.00		-	-	-	-
0.365	-	9.27		1	-	-	M
0.375	3/8	9.52		-	-	10	-
0.394	-	10.00		-	-	-	M6
0.413	-	10.49		0	-	-	-
0.433	-	11.00		-	-	-	-
0.437	7/16	11.11		-	-	1/4 & 12	-
0.445	-	11.30		-	1/4	-	-
0.472	-	12.00		-	-	-	-
0.500	1/2	12.70		-	-	5/16	-
0.512	-	13.00		-	-	-	M8
0.525	-	13.33		-	5/16	-	-
0.551	-	14.00		-	-	-	-
0.562	9/16	14.29		-	-	3/8	-
0.591	-	15.00		-	-	-	-
0.600	-	15.24		-	3/8	-	-
0.625	5/8	15.87		-	-	7/16B	-
0.630	-	16.00		-	-	-	-

Spanner size (AF)				Thread			
Decimal	Fractions	Metric					
in	in	mm		BA	Whit. /BSF	Unified	Metric
0.669	-	17.00		-	-	-	M10
0.687	11/16	17.46		-	-	7/16N	-
0.709	-	18.00		-	-	-	-
0.710	-	18.03		-	7/16	-	-
0.748	-	19.00		-	-	-	M12
0.750	3/4	19.05		-	-	1/2	-
0.787	-	20.00		-	-	-	--
0.812	13/16	20.64		-	-	9/16B	-
0.820	-	20.83		-	1/2	-	-
0.827	-	21.00		-	-	-	-
0.866	-	22.00		-	-	-	M14
0.875	7/8	22.22		-	-	9/16N	-
0.906	-	23.00		-	-	-	-
0.920	-	23.37		-	9/16	-	-
0.937	15/16	23.81		-	-	5/8	-
0.945	-	24.00		-	-	-	M16
0.984	-	25.00		-	-	-	-
1.000	1	25.40		-	-	-	-

AF = Across flat
BA = British Association thread
BSF= British Standard fine thread

CHAPTER 8
THREAD DATA: INCLUDING TAPPING DRILL SIZES

SCREW THREAD TABLES

Full details are given in Tables 8.1 – 8.15. These tables should be studied carefully in order to gain maximum benefit from their use. The thread forms are illustrated in Chapter 9.

NOTES

The tables include the following information for each thread size (see column 1).

1. Basic thread data: These include outside diameter, core diameter and pitch (columns 2–4).

2. Tapping drills, size and depth of thread: The details relating to tapping drill size and depth of thread (columns 5 & 6) are intended to provide readers with the maximum amount of information on which to base their final decision regarding the size of drill to use.

2.1. A tapping drill size may be chosen to give a full (100%) thread depth, but little more will be achieved in terms of strength than one with around 75% maximum thread depth. However, there will be a considerable increase in the amount of torque that needs to be applied to the tap, thereby increasing the likelihood of a broken tap.

2.2. Depth of thread (seldom included in tables of tapping drill sizes) gives an indication of the effect of choosing different-sized drills and of what is being, or can be, achieved.

2.3. The figures given in these tables are aimed at achieving a thread depth of between 60% (for smaller sizes) and 70% (for larger sizes).

3. Diameters: The diameters required to give 60% and 70% thread depths (columns 7 & 8) are given in order to assist the reader when it comes to choosing a smaller or a larger drill for a greater or smaller thread depth, as appropriate for the application.

3.1 The figures show that, at the smaller thread sizes, going up or down a drill size will considerably alter the percentage thread achieved. This highlights that, at these sizes, it is very important that the drill does not drill over size.

4. Conversions: The columns on the right (columns 9-12) give the metric or imperial equivalents of the most important parameters listed on the left (columns 1–8). These will enable readers to choose imperial or metric size drills.

5. Miscellanous:

5.1. The following advice indicates the considerations that affect the final choice of tapping drill:
- For smaller taps, harder materials and deeper holes, tend towards larger drills.

- For stronger threads, larger taps, softer materials and holes in thin sheet, tend toward smaller drills.

5.2. The ability to accept a reduced internal thread depth is based on the assumption that the external thread will be complete and not drastically undersize.

5.3. The percentage is based on the proportion of the difference between the outside diameter and the core diameter and the difference between the outside diameter and the tapping drill diameter.

Table 8.1. ISO metric threads: coarse

Size and dimensions								Conversions (to imperial)				
Size	Out. diam.	Core dia.	Pitch	Tap drill	Depth thread	Diam. 60%	Diam. 70%		Out. diam.	Pitch	Dia. 60%	Dia. 70%
	mm	mm	mm	mm	%	mm	mm		in	TPI	in	in
M1	1.0	0.69	0.25	0.80	65	0.82	0.79		0.039	101.60	0.032	0.031
M1.1*	1.1	0.79	0.25	0.90	65	0.92	0.89		0.043	101.60	0.036	0.035
M1.2	1.2	0.89	0.25	1.00	65	1.02	0.99		0.047	101.60	0.040	0.039
M1.4*	1.4	1.03	0.30	1.15	68	1.18	1.14		0.055	84.67	0.046	0.045
M1.6	1.6	1.17	0.35	1.35	58	1.34	1.30		0.063	72.57	0.053	0.051
M1.8*	1.8	1.37	0.35	1.55	58	1.54	1.50		0.071	72.57	0.061	0.059
M2	2.0	1.51	0.40	1.70	61	1.71	1.66		0.079	63.50	0.067	0.065
M2.2*	2.2	1.65	0.45	1.85	63	1.87	1.81		0.087	56.44	0.074	0.071
M2.5	2.5	1.95	0.45	2.15	63	2.17	2.11		0.098	56.44	0.085	0.083
M3	3.0	2.39	0.50	2.60	65	2.63	2.57		0.118	50.80	0.104	0.101
M3.5*	3.5	2.76	0.60	3.00	68	3.06	2.98		0.138	42.33	0.120	0.118
M4	4.0	3.14	0.70	3.40	70	3.48	3.40		0.157	36.29	0.137	0.134
M4.5*	4.5	3.58	0.75	3.90	65	3.95	3.86		0.177	33.87	0.155	0.152
M5	5.0	4.02	0.80	4.40	61	4.41	4.31		0.197	31.75	0.174	0.170
M6	6.0	4.77	1.00	5.20	65	5.26	5.14		0.236	25.40	0.207	0.202
M7*	7.0	5.77	1.00	6.10	73	6.26	6.14		0.276	25.40	0.247	0.242
M8	8.0	6.47	1.25	6.90	72	7.08	6.93		0.315	20.32	0.279	0.273
M9*	9.0	7.47	1.25	7.90	72	8.08	7.93		0.354	20.32	0.318	0.312
M10	10.0	8.16	1.50	8.70	71	8.90	8.71		0.394	16.93	0.350	0.343
M11*	11.0	9.16	1.50	9.70	71	9.90	9.71		0.433	16.93	0.390	0.382
M12	12.0	9.85	1.75	10.50	70	10.71	10.50		0.472	14.51	0.422	0.413
M14*	14.0	11.55	2.00	12.30	69	12.53	12.28		0.551	12.70	0.493	0.484
M16	16.0	13.55	2.00	14.25	71	14.53	14.28		0.630	12.70	0.572	0.562
M18*	18.0	14.93	2.50	15.75	73	16.16	15.85		0.709	10.16	0.636	0.624
M20	20.0	16.93	2.50	17.75	73	18.16	17.85		0.787	10.16	0.715	0.703
M22*	22.0	18.93	2.50	19.75	73	20.16	19.85		0.866	10.16	0.794	0.782
M24	24.0	20.32	3.00	21.25	75	21.79	21.42		0.945	8.47	0.858	0.843

Non-preferred sizes

Table 8.2. ISO metric threads: fine

Size and dimensions								Conversions (to imperial)			
Size	Out. dia.	Core dia.	Pitch	Tap drill	Depth	Dia. 60%	Dia. 70%	Out. dia.	Pitch	Dia. 60%	Dia. 70%
	mm	mm	mm	mm	%	mm	mm	in	TPI	in	in
M1 × 0.20	1.0	0.75	0.20	0.85	61	0.85	0.83	0.039	127.00	0.034	0.033
M1.2 × 0.20	1.2	0.95	0.20	1.05	61	1.05	1.03	0.047	127.00	0.041	0.040
M1.4 × 0.20*	1.4	1.15	0.20	1.25	61	1.25	1.23	0.055	127.00	0.049	0.048
M1.6 × 0.20	1.6	1.35	0.20	1.45	61	1.45	1.43	0.063	127.00	0.057	0.056
M1.8 × 0.20*	1.8	1.55	0.20	1.65	61	1.65	1.63	0.071	127.00	0.065	0.064
M2 × 0.25	2.0	1.69	0.25	1.80	65	1.82	1.79	0.079	101.60	0.071	0.070
M2.2 × 0.25*	2.2	1.89	0.25	2.00	65	2.02	1.99	0.087	101.60	0.079	0.078
M2.5 × 0.35	2.5	2.07	0.35	2.25	58	2.24	2.20	0.098	72.57	0.088	0.087
M3 × 0.35	3.0	2.57	0.35	2.75	58	2.74	2.70	0.118	72.57	0.108	0.106
M3.5 × 0.35*	3.5	3.07	0.35	3.20	70	3.24	3.20	0.138	72.57	0.128	0.126
M4 × 0.50	4.0	3.39	0.50	3.60	65	3.63	3.57	0.157	50.80	0.143	0.141
M4.5 × 0.50*	4.5	3.89	0.50	4.10	65	4.13	4.07	0.177	50.80	0.163	0.160
M5 × 0.50	5.0	4.39	0.50	4.60	65	4.63	4.57	0.197	50.80	0.182	0.180
M6 × 0.75	6.0	5.08	0.75	5.40	65	5.45	5.36	0.236	33.87	0.214	0.211
M7 × 0.75*	7.0	6.08	0.75	6.40	65	6.45	6.36	0.276	33.87	0.254	0.250
M8 × 0.75	8.0	7.08	0.75	7.40	65	7.45	7.36	0.315	33.87	0.293	0.290
M8 × 1.00	8.0	6.77	1.00	7.20	65	7.26	7.14	0.315	25.40	0.286	0.281
M9 × 0.75*	9.0	8.08	0.75	8.40	65	8.45	8.36	0.354	33.87	0.333	0.329
M9 × 1.00*	9.0	7.77	1.00	8.20	65	8.26	8.14	0.354	25.40	0.325	0.321
M10 × 1.25	10.0	8.47	1.25	8.90	72	9.08	8.93	0.394	20.32	0.357	0.351
M11 × 1.00*	11.0	9.77	1.00	10.10	73	10.26	10.14	0.433	25.40	0.404	0.399
M12 × 1.00	12.0	10.77	1.00	11.10	73	11.26	11.14	0.472	25.40	0.443	0.439
M12 × 1.50	12.0	10.16	1.50	10.70	71	10.90	10.71	0.472	16.93	0.429	0.422
M14 × 1.00*	14.0	12.77	1.00	13.10	73	13.26	13.14	0.551	25.40	0.522	0.517
M14 × 1.50*	14.0	12.16	1.50	12.70	71	12.90	12.71	0.551	16.93	0.508	0.500
M16 × 1.00	16.0	14.77	1.00	15.00	82	15.26	15.14	0.630	25.40	0.601	0.596
M18 × 1.00	18.0	16.77	1.00	17.00	82	17.26	17.14	0.709	25.40	0.680	0.675
M18 × 2.00*	18.0	15.55	2.00	16.00	82	16.53	16.28	0.709	12.70	0.651	0.641
M20 × 1.00	20.0	18.77	1.00	19.00	82	19.26	19.14	0.787	25.40	0.758	0.754
M20 × 2.00	20.0	17.55	2.00	18.00	82	18.53	18.28	0.787	12.70	0.729	0.720
M22 × 1.00*	22.0	20.77	1.00	21.00	82	21.26	21.14	0.866	25.40	0.837	0.832
M22 × 1.50*	22.0	20.16	1.50	20.50	82	20.90	20.71	0.866	16.93	0.823	0.815
M22 × 2.00*	22.0	19.55	2.00	20.00	82	20.53	20.28	0.866	12.70	0.808	0.799
M24 × 1.00	24.0	22.77	1.00	23.00	82	23.26	23.14	0.945	25.40	0.916	0.911
M24 × 1.50	24.0	22.16	1.50	22.50	82	22.90	22.71	0.945	16.93	0.901	0.894
M24 × 2.00	24.0	21.55	2.00	22.00	82	22.53	22.28	0.945	12.70	0.887	0.877

*Non-preferred sizes

Note: See also Table 8.3 (ISO metric threads: spark plugs) and Table 8.4 (ISO metric threads: conduits)

Table 8.3 ISO metric threads: spark plugs

Size and dimensions									Conversions (to imperial)			
Size	Out. dia.	Core dia.	Pitch	Tap drill	Depth	Dia. 60%	Dia. 70%		Dia.	Pitch	Dia. 60%	Dia. 70%
	mm	mm	mm	mm	mm	mm	mm		in	TPI	in	in
M10 × 1.0	10.0	8.77	1.00	9.20	65	9.26	9.14		0.394	25.40	0.365	0.360
M12 × 1.25	12.0	10.47	1.25	10.90	72	11.08	10.93		0.472	20.32	0.436	0.430
M14 × 1.25	14.0	12.47	1.25	12.90	72	13.08	12.93		0.551	20.32	0.515	0.509
M18 × 1.5	18.0	16.16	1.50	16.75	68	16.90	16.71		0.709	16.93	0.665	0.658

Table 8.4 ISO metric threads: conduit

Size and dimensions									Conversions (to imperial)			
Size	Out. dia.	Core dia.	Pitch	Tap drill	Depth	Dia. 60%	Dia. 70%		Dia.	Pitch	Dia. 60%	Dia. 70%
	mm	mm	mm	mm	mm	mm	mm		in	TPI	in	in
M16 × 1.5	16.0	14.160	1.50	14.50	82	14.90	14.71		0.630	16.93	0.586	0.579
M20 × 1.5	20.0	18.160	1.50	18.50	82	18.90	18.71		0.787	16.93	0.744	0.737
M25 × 1.5	25.0	23.160	1.50	23.50	82	23.90	23.71		0.984	16.93	0.941	0.934
M32 × 1.5	32.0	30.160	1.50	30.50	82	30.90	30.71		1.260	16.93	1.216	1.209
M40 × 1.5	40.0	38.160	1.50	38.50	82	38.90	38.71		1.575	16.93	1.531	1.524

Table 8.5. British Association (BA) threads

Size and dimensions									Conversion (to imperial)			
Size	Out. dia.	Core dia.	Pitch	Tap drill	Depth	Dia. 60%	Dia. 70%		Out. dia.	Pitch	Dia. 60%	Dia. 70%
	mm	mm	mm	mm	mm	mm	mm		in	TPI	in	in
0	6.00	4.80	1.00	5.20	67	5.28	5.16		0.236	25.40	0.208	0.203
1	5.30	4.22	0.90	4.60	65	4.65	4.54		0.209	28.22	0.183	0.179
2	4.70	3.73	0.81	4.00	72	4.12	4.02		0.185	31.36	0.162	0.158
3	4.10	3.22	0.73	3.50	68	3.57	3.49		0.161	34.79	0.141	0.137
4	3.60	2.81	0.66	3.10	63	3.12	3.05		0.142	38.48	0.123	0.120
5	3.20	2.49	0.59	2.75	64	2.78	2.70		0.126	43.05	0.109	0.106
6	2.80	2.16	0.53	2.40	63	2.42	2.35		0.110	47.92	0.095	0.093
7	2.50	1.92	0.48	2.10	69	2.15	2.10		0.098	52.92	0.085	0.083
8	2.20	1.68	0.43	1.85	68	1.89	1.84		0.087	59.07	0.074	0.072
9	1.90	1.43	0.39	1.60	64	1.62	1.57		0.075	65.13	0.064	0.062
10	1.70	1.28	0.35	1.45	60	1.45	1.41		0.067	72.57	0.057	0.055
11	1.50	1.13	0.31	1.25	67	1.28	1.24		0.059	81.94	0.050	0.049
12	1.30	0.96	0.28	1.10	60	1.10	1.06		0.051	90.71	0.043	0.042
13	1.20	0.90	0.25	1.00	67	1.02	0.99		0.047	101.60	0.040	0.039
14	1.00	0.72	0.23	0.82	65	0.83	0.81		0.039	110.43	0.033	0.032
15	0.90	0.65	0.21	0.75	60	0.75	0.72		0.035	120.95	0.029	0.028
16	0.79	0.56	0.19	0.65	61	0.65	0.63		0.031	133.68	0.026	0.025
17	0.70	0.50	0.17	0.58	59	0.58	0.56		0.028	149.41	0.023	0.022
18	0.62	0.44	0.15	0.50	67	0.51	0.49		0.024	169.33	0.020	0.019

Table 8.6. Unified Threads: coarse (UNC)

| Size and dimensions | | | | | | | | Conversions (metric and imperial) | | | |
Size No./ Inch	Out. dia.	Core dia.	Pitch	Tap drill	Depth (%)	Dia. 60%	Dia. 70%	Dia.	Pitch	Dia. 60%	Dia. 70%
	in	in	TPI	mm	%	mm	mm	mm	mm	in	in
No. 1	0.0730	0.054	64	1.55	62	1.56	1.51	1.85	0.40	0.061	0.060
No. 2	0.0860	0.064	56	1.85	60	1.85	1.79	2.18	0.45	0.073	0.071
No. 3	0.0990	0.073	48	2.10	64	2.13	2.06	2.51	0.53	0.084	0.081
No. 4	0.1120	0.081	40	2.35	64	2.38	2.30	2.84	0.64	0.094	0.091
No. 5	0.1250	0.094	40	2.65	67	2.71	2.63	3.17	0.64	0.107	0.104
No. 6	0.1380	0.100	32	2.85	67	2.92	2.82	3.51	0.79	0.115	0.111
No. 8	0.1640	0.126	32	3.50	68	3.58	3.48	4.17	0.79	0.141	0.137
No. 10	0.1900	0.139	24	3.90	71	4.05	3.92	4.83	1.06	0.159	0.154
No. 12	0.2160	0.165	24	4.50	76	4.71	4.58	5.49	1.06	0.185	0.180
1/4in	0.2500	0.189	20	5.10	80	5.42	5.26	6.35	1.27	0.213	0.207
5/16in	0.3125	0.244	18	6.60	77	6.90	6.73	7.94	1.41	0.272	0.265
3/8in	0.3750	0.298	16	8.00	78	8.36	8.16	9.52	1.59	0.329	0.321
7/16in	0.4375	0.350	14	9.40	77	9.78	9.55	11.11	1.81	0.385	0.376
1/2in	0.5000	0.406	13	10.80	79	11.26	11.02	12.70	1.95	0.443	0.434
9/16in	0.5625	0.460	12	12.20	80	12.73	12.47	14.29	2.12	0.501	0.491
5/8in	0.6250	0.513	11	13.50	84	14.18	13.89	15.87	2.31	0.558	0.547
3/4in	0.7500	0.627	10	16.50	82	17.18	16.87	19.05	2.54	0.676	0.664
7/8	0.8750	0.739	9	19.50	79	20.15	19.80	22.22	2.82	0.793	0.780
1in	1.0000	0.847	8	22.50	74	23.06	22.67	25.40	3.17	0.908	0.893
1 1/8in	1.1250	0.950	7	25.00	80	25.90	25.46	28.57	3.63	1.020	1.002

Table 8.7. Unified Threads: fine (UNF)

| Size and dimensions | | | | | | | | Conversions (metric and imperial | | | |
Size No./In.	Out. dia.	Core dia.	Pitch	Tap drill	Depth	Dia. 60%	Dia. 70%	Dia.	Pitch	Dia. 60%	Dia. 70%
	in	in	TPI	mm	%	mm	mm	mm	mm	in	in
No. 0	0.0600	0.045	80	1.30	58	1.29	1.25	1.52	0.32	0.051	0.049
No. 1	0.0730	0.056	72	1.60	59	1.59	1.55	1.85	0.35	0.063	0.061
No. 2	0.0860	0.067	64	1.90	58	1.89	1.84	2.18	0.40	0.074	0.073
No. 3	0.0990	0.077	56	2.20	57	2.18	2.13	2.51	0.45	0.086	0.084
No. 4	0.1120	0.086	48	2.45	61	2.46	2.39	2.84	0.53	0.097	0.094
No. 5	0.1250	0.097	44	2.70	67	2.75	2.68	3.17	0.58	0.108	0.105
No. 6	0.1380	0.107	40	3.00	65	3.04	2.96	3.51	0.64	0.120	0.117
No. 8	0.1640	0.130	36	3.60	65	3.65	3.56	4.17	0.71	0.144	0.140
No. 10	0.1900	0.152	32	4.20	64	4.24	4.14	4.83	0.79	0.167	0.163
No. 12	0.2160	0.172	28	4.80	62	4.82	4.71	5.49	0.91	0.190	0.185
1/4in	0.2500	0.206	28	5.60	67	5.68	5.57	6.35	0.91	0.224	0.219
5/16in	0.3125	0.261	24	7.10	65	7.16	7.03	7.94	1.06	0.282	0.277
3/8in	0.3750	0.324	24	8.60	71	8.75	8.62	9.52	1.06	0.344	0.339
7/16in	0.4375	0.376	20	10.00	71	10.18	10.02	11.11	1.27	0.401	0.395
1/2in	0.5000	0.439	20	11.60	71	11.77	11.61	12.70	1.27	0.463	0.457
9/16in	0.5625	0.494	18	13.00	74	13.25	13.08	14.29	1.41	0.522	0.515
5/8in	0.6250	0.557	18	14.50	79	14.84	14.66	15.87	1.41	0.584	0.577
3/4in	0.7500	0.673	16	17.75	67	17.88	17.69	19.05	1.59	0.704	0.696
7/8in	0.8750	0.787	14	20.75	66	20.89	20.67	22.22	1.81	0.822	0.814
1in	1.0000	0.898	12	23.50	73	23.84	23.58	25.40	2.12	0.939	0.928

Table 8.8. British Standard Whitworth threads (BSW)

	Size and dimensions							Conversions (metric and imperial)			
Size	Out. dia.	Core dia.	Pit ch	Tap drill	Depth	Dia. 60%	Dia. 70%	Out. dia.	Pitch	Dia. 60%	Dia. 70%
	in	in	TPI	mm	%	mm	mm	mm	mm	in	in
1/16	0.0625	0.041	60	1.25	62	1.26	1.21	1.59	0.42	0.050	0.048
3/32	0.0938	0.067	48	1.95	64	1.98	1.91	2.38	0.53	0.078	0.075
1/8	0.1250	0.093	40	2.65	65	2.69	2.61	3.17	0.64	0.106	0.103
5/32	0.1562	0.116	32	3.30	66	3.36	3.26	3.97	0.79	0.132	0.128
3/16	0.1875	0.134	24	3.90	64	3.95	3.81	4.76	1.06	0.155	0.150
7/32	0.2188	0.165	24	4.70	63	4.74	4.61	5.56	1.06	0.187	0.181
1/4	0.2500	0.186	20	5.30	65	5.37	5.21	6.35	1.27	0.212	0.205
5/16	0.3125	0.241	18	6.80	63	6.85	6.67	7.94	1.41	0.270	0.263
3/8	0.3750	0.295	16	8.10	70	8.31	8.10	9.52	1.59	0.327	0.319
7/16	0.4375	0.346	14	9.50	69	9.72	9.49	11.11	1.81	0.383	0.373
1/2	0.5000	0.393	12	10.80	70	11.07	10.80	12.70	2.12	0.436	0.425
9/16	0.5625	0.456	12	12.40	70	12.66	12.39	14.29	2.12	0.498	0.488
5/8	0.6250	0.509	11	13.70	74	14.10	13.80	15.87	2.31	0.555	0.544
11/16	0.6875	0.571	11	15.25	75	15.69	15.39	17.46	2.31	0.618	0.606
3/4	0.7500	0.622	10	16.75	71	17.10	16.77	19.05	2.54	0.673	0.660
13/16	0.8125	0.684	10	18.25	73	18.69	18.36	20.64	2.54	0.736	0.723
7/8	0.8750	0.733	9	19.50	75	20.06	19.69	22.22	2.82	0.790	0.775
15/16	0.9375	0.795	9	21.25	71	21.64	21.28	23.81	2.82	0.852	0.838
1	1.0000	0.840	8	22.50	71	22.96	22.55	25.40	3.17	0.904	0.888
1 1/8	1.1250	0.942	7	25.25	72	25.79	25.32	28.57	3.63	1.015	0.997

Table 8.9. British Standard Fine (BSF) threads (Whitworth Form)

	Size and dimensions							Conversions (metric and imperial)			
Size	Out. dia.	Core dia.	Pitch	Tap drill	Depth	Dia. 60%	Dia. 70%	Out. dia.	Pitch		
	in	in	TPI	mm	%	mm	mm	mm	mm	in	in
3/16	0.1875	0.147	32	4.10	65	4.15	4.05	4.76	0.79	0.163	0.159
7/32	0.2188	0.173	28	4.80	65	4.86	4.74	5.56	0.91	0.191	0.187
1/4	0.2500	0.201	26	5.50	68	5.60	5.47	6.35	0.98	0.220	0.216
5/16	0.3125	0.254	22	6.90	70	7.05	6.90	7.94	1.15	0.278	0.272
3/8	0.3750	0.311	20	8.40	69	8.55	8.39	9.52	1.27	0.337	0.330
7/16	0.4375	0.366	18	9.90	67	10.03	9.85	11.11	1.41	0.395	0.388
1/2	0.5000	0.420	16	11.30	69	11.48	11.28	12.70	1.59	0.452	0.444
9/16	0.5625	0.482	16	12.90	68	13.07	12.86	14.29	1.59	0.514	0.506
5/8	0.6250	0.534	14	14.25	70	14.48	14.25	15.87	1.81	0.570	0.561
11/16	0.6875	0.596	14	15.75	74	16.07	15.84	17.46	1.81	0.633	0.623
3/4	0.7500	0.643	12	17.00	76	17.42	17.15	19.05	2.12	0.686	0.675
7/8	0.8750	0.759	11	20.00	75	20.45	20.15	22.22	2.31	0.805	0.794
1	1.0000	0.872	10	23.00	74	23.45	23.12	25.40	2.54	0.923	0.910
1 1/8	1.1250	0.983	9	25.75	78	26.41	26.04	28.57	2.82	1.040	1.025
1 1/4	1.2500	1.108	9	29.00	76	29.58	29.22	31.75	2.82	1.165	1.150

Table 8.10. Model engineer thread (Whitworth form)

Size and dimensions								Conversions (metric and imperial)			
Size	Out. dia.	Core dia.	Pitch	Tap drill	Depth	Dia. 60%	Dia. 70%	Out. dia.	Pitch	Dia. 60%	Dia. 70%
	in	in	TPI	mm	%	mm	mm	mm	mm	in	in
1/8	0.1250	0.093	40	2.65	65	2.69	2.61	3.17	0.64	0.106	0.103
5/32	0.1562	0.124	40	3.40	70	3.48	3.40	3.97	0.64	0.137	0.134
3/16	0.1875	0.155	40	4.20	69	4.27	4.19	4.76	0.64	0.168	0.165
7/32	0.2188	0.187	40	5.00	68	5.07	4.99	5.56	0.64	0.200	0.196
1/4	0.2500	0.218	40	5.80	68	5.86	5.78	6.35	0.64	0.231	0.228
5/16	0.3125	0.280	40	7.30	78	7.45	7.37	7.94	0.64	0.293	0.290
3/8	0.3750	0.343	40	8.90	77	9.04	8.96	9.52	0.64	0.356	0.353
1/4	0.2500	0.210	32	5.70	64	5.74	5.64	6.35	0.79	0.226	0.222
9/32	0.2813	0.241	32	6.50	63	6.53	6.43	7.14	0.79	0.257	0.253
5/16	0.3125	0.272	32	7.20	73	7.33	7.23	7.94	0.79	0.288	0.284
3/8	0.3750	0.335	32	8.80	71	8.92	8.81	9.52	0.79	0.351	0.347
7/16	0.4375	0.397	32	10.40	70	10.50	10.40	11.11	0.79	0.413	0.409
1/2	0.5000	0.460	32	12.00	69	12.09	11.99	12.70	0.79	0.476	0.472

Table 8.11. British Standard Brass thread 26TPI (Whitworth form)

Size and dimensions								Conversions (metric and imperial)			
Size	Out. dia.	Core dia.	Pitch	Tap drill	Depth	Dia. 60%	Dia. 70%	Out. dia.	Pitch	Dia. 60%	Dia. 70%
	in	in	TPI	mm	%	mm	mm	mm	mm	in	in
1/4	0.2500	0.201	26	5.50	68	5.60	5.47	6.35	0.98	0.220	0.216
5/16	0.3125	0.263	26	7.10	67	7.19	7.06	7.94	0.98	0.283	0.278
3/8	0.3750	0.326	26	8.70	66	8.77	8.65	9.52	0.98	0.345	0.341
7/16	0.4375	0.388	26	10.20	73	10.36	10.24	11.11	0.98	0.408	0.403
1/2	0.5000	0.451	26	11.80	72	11.95	11.82	12.70	0.98	0.470	0.466
5/8	0.6250	0.576	26	15.00	70	15.12	15.00	15.87	0.98	0.595	0.591
3/4	0.7500	0.701	26	18.25	64	18.30	18.17	19.05	0.98	0.720	0.716

Table 8.12. British Standard Cycle thread 26TPI

Size and dimensions								Conversions (metric and imperial)			
Size	Out. dia.	Core dia.	Pitch	Tap drill	Depth	Dia. 60%	Dia. 70%	Out. dia.	Pitch	Dia. 60%	Dia. 70%
	in	in	TPI	mm	%	mm	mm	mm	mm	in	in
1/4	0.2500	0.209	26	5.70	62	5.73	5.62	6.35	0.98	0.225	0.221
5/16	0.3125	0.272	26	7.30	61	7.31	7.21	7.94	0.98	0.288	0.284
3/8	0.3750	0.334	26	8.80	70	8.90	8.80	9.52	0.98	0.350	0.346
7/16	0.4375	0.397	26	10.40	68	10.49	10.38	11.11	0.98	0.413	0.409
1/2	0.5000	0.459	26	12.00	67	12.08	11.97	12.70	0.98	0.475	0.471
9/16	0.5625	0.522	26	13.60	66	13.66	13.56	14.29	0.98	0.538	0.534
5/8	0.6250	0.584	26	15.25	60	15.25	15.15	15.87	0.98	0.600	0.596
11/16	0.6875	0.647	26	16.75	68	16.84	16.73	17.46	0.98	0.663	0.659
3/4	0.7500	0.709	26	18.25	77	18.43	18.32	19.05	0.98	0.725	0.721

Table 8.13. British Standard Pipe Parallel (RP series) (Whitworth form)

Size and dimensions					Conversions (metric and imperial)		
Nominal size (NS)*	Pipe OD	Pitch	Tapping drill		Pipe OD	Pitch	Tapping drill
	in	TPI	mm		mm	mm	in
1/8	0.383	28.0	8.60		9.73	0.91	0.339
1/4	0.518	19.0	11.50		13.16	1.34	0.453
3/8	0.656	19.0	15.00		16.66	1.34	0.591
1/2	0.825	14.0	18.50		20.95	1.81	0.728
3/4	1.041	14.0	24.00		26.44	1.81	0.945
1	1.309	11.0	30.25		33.25	2.31	1.191
1 1/4	1.650	11.0	39.00		41.91	2.31	1.535
1 1/2	1.882	11.0	45.00		47.80	2.31	1.772
2	2.347	11.0	56.50		59.61	2.31	2.224

Note that the actual outside diameters are larger than the nominal size (NS). The amount varies from about +¼in at the smaller sizes to about +⅜in at the larger.

Table 8.14. British Standard Pipe Taper (BSPT) threads: RC Series

Size and dimensions						Conversions (metric and imperial)			
Nominal size (ns)	Pipe OD	Pitch	Tapping drill (WR)	Tapping drill (WOR)		Pipe OD	Pitch	Tapping drill (WR)	Tapping drill (WOR)
	in	TPI	mm	mm		mm	mm	in	in
1/8	0.383	28.0	8.00	8.40		9.73	0.91	0.315	0.331
1/4	0.518	19.0	10.80	11.20		13.16	1.34	0.425	0.441
3/8	0.656	19.0	14.25	14.75		16.66	1.34	0.561	0.581
1/2	0.825	14.0	17.75	18.25		20.95	1.81	0.699	0.719
3/4	1.041	14.0	23.00	23.75		26.44	1.81	0.906	0.935
1	1.309	11.0	29.00	30.00		33.25	2.31	1.142	1.181
1 1/4	1.650	11.0	37.50	38.50		41.91	2.31	1.476	1.516
1 1/2	1.882	11.0	43.50	44.50		47.80	2.31	1.713	1.752
2	2.347	11.0	55.00	56.00		59.61	2.31	2.165	2.205
2 1/2	2.960	11.0	70.00	71.00		75.18	2.31	2.756	2.795

WR   With taper reamer (recommended),

WOR   Without taper reamer,

NS   Note that the actual outside diameters are larger than the nominal size, varying from about ¼in at the smaller sizes to about ⅜in at the larger sizes.

Note: In view of the demanding duties that these threads are called to perform, e.g. as joints in high pressure systems, made without sealers, it is essential that good-quality threads are made and that the NPT and BSPT systems are not intermixed.

Table 8.15. National Pipe Taper (NPT) threads

| Size and dimensions | | | | | | Conversions (metric and imperial) | | | |
Nominal size (ns	Pipe OD	Pitch	Tapping drill (WR)	Tapping drill (WOR)		Pipe OD	Pitch	Tapping drill (WR)	Tapping drill (WOR)
	in	TPI	mm	mm		mm	mm	in	in
1/16	0.312	27.0	6.00	6.30		7.92	0.94	0.236	0.248
1/8	0.405	27.0	8.40	8.70		10.29	0.94	0.331	0.343
1/4	0.540	18.0	10.70	11.10		13.72	1.41	0.421	0.437
3/8	0.675	18.0	14.25	14.50		17.14	1.41	0.561	0.571
1/2	0.840	14.0	17.50	18.00		21.34	1.81	0.689	0.709
3/4	1.050	14.0	22.75	23.25		26.67	1.81	0.896	0.915
1	1.315	11.5	28.50	29.00		33.40	2.21	1.122	1.142
1 1/4	1.660	11.5	37.50	38.00		42.16	2.21	1.476	1.496
1 1/2	1.900	11.5	43.50	44.00		48.26	2.21	1.713	1.732
2	2.375	11.5	55.00	56.00		60.32	2.21	2.165	2.205
2 1/2	2.875	8.0	66.00	67.00		73.02	3.17	2.598	2.638

Notes: See Table 8.14 (BSPT threads).

MODEL ENGINEER METRIC THREADS

In the metric system no standard exists for threads for model engineering, but a group of model engineers have combined to recommend selected metric threads from other standards for model engineering projects. The fasteners though have hexagons which are considered unsuitable for accurate scale modelling. The recommendation therefore suggests the use of other hexagon sizes.

There are two ranges, one based on the constant pitch series, with pitches of 0.5mm, 0.75mm and 1.0mm, making them comparable to the fine pitch imperial ME threads and the other on the metric coarse threads.

Table 8.16 Standard ISO metric threads for possible use in model engineering

| Constant pitch ranges | | | Mod. Eng. hexagon | |
Size	Out. dia.	Pitch		
	mm	mm	mm	in.
M3 × 0.50	3.0	0.50	4.50	0.177
M4 × 0.50	4.0	0.50	5.50	0.217
M4.5 × 0.50	4.5	0.50	6.00	0.236
M5 × 0.50	5.0	0.50	7.00	0.276
M5.5 × 0.50	5.5	0.50	7.00	0.276
M6 × 0.50	6.0	0.50	8.00	0.315
M4.5 × 0.75	4.5	0.75	6.00	0.236
M6 × 0.75	6.0	0.75	8.00	0.315
M7 × 0.75	7.0	0.75	9.00	0.354
M8 × 0.75	8.0	0.75	10.00	0.394
M10 × 0.75	10.0	0.75	12.00	0.472
M12 × 0.75	12.0	0.75	14.00	0.551
M10 × 1.00	10.0	1.00	12.00	0.472
M12 × 1.00	12.0	1.00	14.00	0.551
M14 × 1.00	14.0	1.00	17.00	0.669
M16 × 1.00	16.0	1.00	20.00	0.787
M18 × 1.00	18.0	1.00	22.00	0.866
M20 × 1.00	20.0	1.00	24.00	0.945

Table 8.17. Tapping drills

Size	Out. dia.	Core dia.	Pitch	Tap drill	Depth	Dia. 60%	Dia. 70%		Conversion to imperial			
									Out. dia.	Pitch	Dia. 60%	Dia. 70%
	mm	mm	mm	mm	%	mm	mm		in	TPI	in	in
M3 × 0.50	3.0	2.39	0.50	2.60	65	2.63	2.57		0.118	50.80	0.104	0.101
M4 × 0.50	4.0	3.39	0.50	3.60	65	3.63	3.57		0.157	50.80	0.143	0.141
M4.5 x 0.50	4.5	3.89	0.50	4.10	65	4.13	4.07		0.177	50.80	0.163	0.160
M5 × 0.50	5.0	4.39	0.50	4.60	65	4.63	4.57		0.197	50.80	0.182	0.180
M5.5 × 0.50	5.5	4.89	0.50	5.10	65	5.13	5.07		0.217	50.80	0.202	0.200
M6 × 0.50	6.0	5.39	0.50	5.60	65	5.63	5.57		0.236	50.80	0.222	0.219
M4.5 × 0.75	4.5	3.58	0.75	3.90	65	3.95	3.86		0.177	33.87	0.155	0.152
M6 × 0.75	6.0	5.08	0.75	5.40	65	5.45	5.36		0.236	33.87	0.214	0.211
M7 × 0.75	7.0	6.08	0.75	6.40	65	6.45	6.36		0.276	33.87	0.254	0.250
M8 × 0.75	8.0	7.08	0.75	7.30	76	7.45	7.36		0.315	33.87	0.293	0.290
M10 × 0.75	10.0	9.08	0.75	9.30	76	9.45	9.36		0.394	33.87	0.372	0.368
M12 × 0.75	12.0	11.08	0.75	11.30	76	11.45	11.36		0.472	33.87	0.451	0.447
M10 × 1.00	10.0	8.77	1.00	9.10	73	9.26	9.14		0.394	25.40	0.365	0.360
M12 × 1.00	12.0	10.77	1.00	11.10	73	11.26	11.14		0.472	25.40	0.443	0.439
M14 x 1.00	14.0	12.77	1.00	13.10	73	13.26	13.14		0.551	25.40	0.522	0.517
M16 ×1.00	16.0	14.77	1.00	15.10	73	15.26	15.14		0.630	25.40	0.601	0.596
M18 × 1.00	18.0	16.77	1.00	17.10	73	17.26	17.14		0.709	25.40	0.680	0.675
M20 × 1.00	20.0	18.77	1.00	19.10	73	19.26	19.14		0.787	25.40	0.758	0.754

Table 8.18. Standard ISO metric threads for possible use in model engineering.

Size	Variable pitch					Conversion to imperial			
	Out. dia.	Pitch	Normal Hexagon	Mod. Eng. Hexagon		Out. dia.	Pitch	Normal hexagon	Mod. Eng. hexagon
	mm	mm	mm	mm		in	TPI	in	in
M1	1.0	0.25	2.50	1.50		0.039	101.60	0.098	0.059
M1.2	1.2	0.25	3.00	2.00		0.047	101.60	0.118	0.079
M1.4	1.4	0.30	3.00	2.50		0.055	84.67	0.118	0.098
M1.6	1.6	0.35	3.20	3.00		0.063	72.57	0.126	0.118
M1.8	1.8	0.35	-	3.20		0.071	72.57	-	0.126
M2	2.0	0.40	4.00	3.50		0.079	63.50	0.157	0.138
M2.2	2.2	0.45	-	4.00		0.087	56.44	-	0.157
M2.5	2.5	0.45	5.00	4.50		0.098	56.44	0.197	0.177
M3	3.0	0.50	5.50	5.00		0.118	50.80	0.217	0.197
M3.5	3.5	0.60	-	6.00		0.138	42.33	-	0.236
M4	4.0	0.70	7.00	7.00		0.157	36.29	0.276	0.276
M4.5	4.5	0.75	-	8.00		0.177	33.87	-	0.315
M5	5.0	0.80	8.00	9.00		0.197	31.75	0.315	0.354
M6	6.0	1.00	10.00	10.00		0.236	25.40	0.394	0.394

See page 39 for tapping drills sizes and other data

FLUTELESS TAPS

Fluteless taps produce a thread by cold-forming so no swarf is produced. However, they only work satisfactorily with ductile materials, e.g. mild steel, aluminum and copper. With care, and some experimentation, more difficult materials, e.g. autensic stainless steel, may possibly be tapped.

The main advantage that fluteless taps have over normal taps is their ability to produce stronger threads, due to the work hardening that results from the cold-forming process. This makes them ideal for tapping holes in thin metal parts. They also produce a swarf-free hole, making for easier assembly.

Having no flutes the taps are stronger, making them less likely to break, even though they require a little more torque to create the thread but as a result, production times in industry are shorter. A good quality lubricant, not a cutting oil, must be used.

TAPPING DRILL SIZE

As the thread is cold-formed rather than cut, the size of tapping drill differs from that required for conventional taps.

The following drill sizes are recommended for use in industrial applications. In some cases, mainly at the larger sizes, a larger drill size may be employed satisfactorily. Some experimentation may be worthwhile.

Table 8.19
ISO Metric Coarse

Thread	Tap drill
No.	mm
M2	1.8
M2.5	2.3
M3	2.8
M3.5	3.2
M4	3.7
M5	4.6
M6	5.6
M7	6.5
M8	7.4
M10	9.3
M12	11.1

Table 8.20
Unified Coarse (UNC)

Thread	Tap drill
No/Inch.	mm
No.2	1.8
No. 4	2.3
No.5	2.8
No.6	3.2
No.8	3.7
No.10	4.3
1/4in	5.8
5/16in	7.3
3/8in	8.8
3/8in	6.5

Table 8.21
Unified Fine (UNF)

Thread	Tap drill
No./ Inch	mm
No. 4	2.6
No. 6	3.2
No. 8	3.8
No. 10	4.5
1/4in	5.9
5/16in	7.5
3/8in	9.0
7/16in	10.4

Table 8.22. British Association (BA)

Thread	Tap drill
No.	mm
0	5.6
2	4.4
4	3.3
6	2.6
8	2.0

CHAPTER 9
SCREW THREAD FORMS

This chapter gives details of the thread forms for the most common metric and imperial thread types.

THREAD FORMS

ISO METRIC AND UNIFIED THREAD

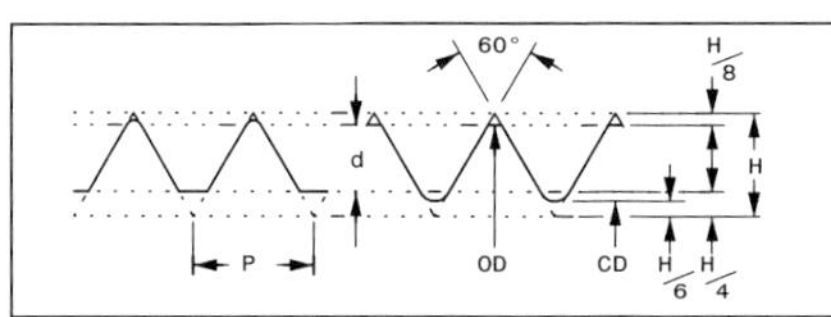

CD   core diameter of screw

d   depth of engagement
     = 0.541266P

H   angular depth
     = 0.866025P

H/4   = 0.216506P

H/6   = 0.144337P

H/8   = 0.108253P

OD   outside diameter of screw

P   pitch

DIFFERENCE BETWEEN THREADS IN NUT AND SCREW

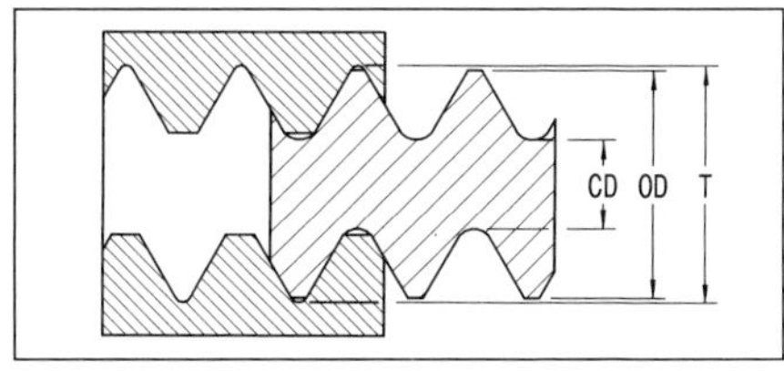

CD   core diameter of screw

OD   outside diameter of screw

T   Note how the radius in the base of the thread in the nut makes the tap larger than the screw diameter.

BRITISH ASSOCIATION THREAD

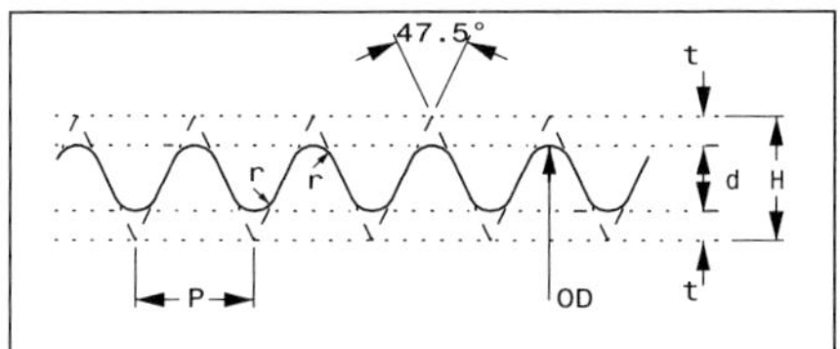

d   depth of thread
     = 0.6P

H   angular depth
     = 1.136P

OD   outside diameter of screw

P   pitch
     = 0.9n mm, where n is the BA number

r   radius
     = 0.18P approx.

t   = 0.27P approx.

BRITISH STANDARD WHITWORTH FORM THREAD

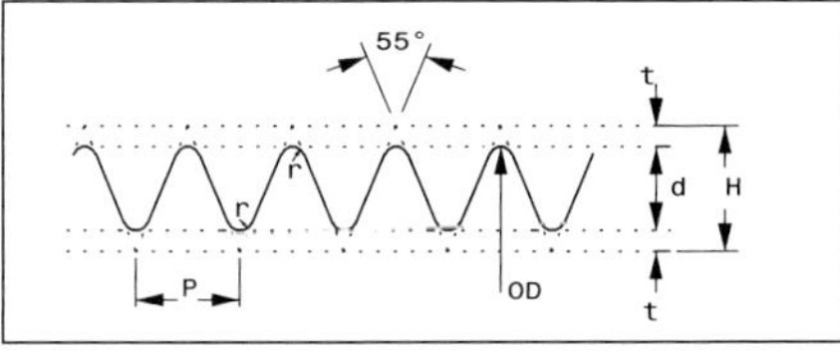

d   depth of thread
     = 0.640327P

= 2/3H
H   angular depth
        = 0.960491P
OD   outside diameter of screw
P   pitch
r   radius
        = 0.137329P approx.
t   = 1/6H

BRITISH STANDARD CYCLE FORM THREAD

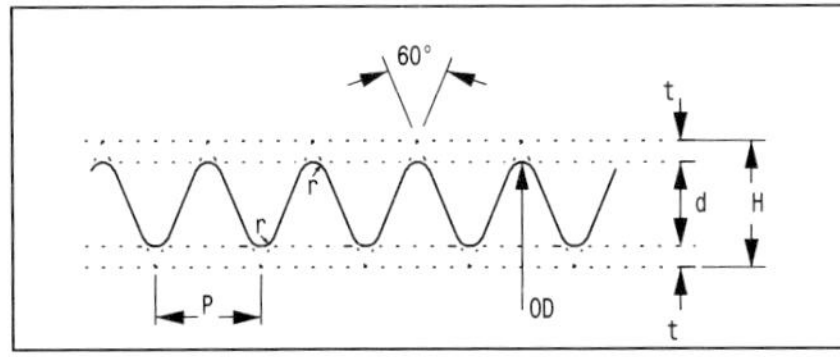

d   depth of thread
        = 0.5327P
        = 2/3H
H   angular depth
        = 0.866P
OD   outside diameter of screw
P   pitch
r   radius
        = 1/6P
t   = 1/6H

BRITISH STANDARD PIPE TAPER THREAD

H   peak-to-peak height
        = 0.9605P

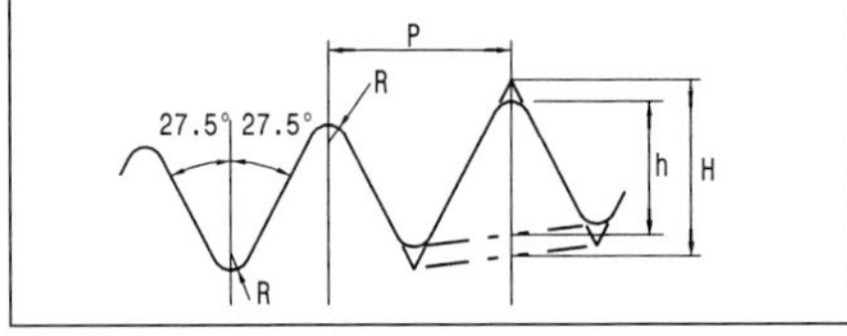

h   height of thread
        = 0.6403P
R   crest and valley radius
        = 0.1371P

Taper   1 in 16 on diameter

NATIONAL PIPE TAPER THREAD

C   truncation, minimum:
        = 0.033P
        truncation, maximum:

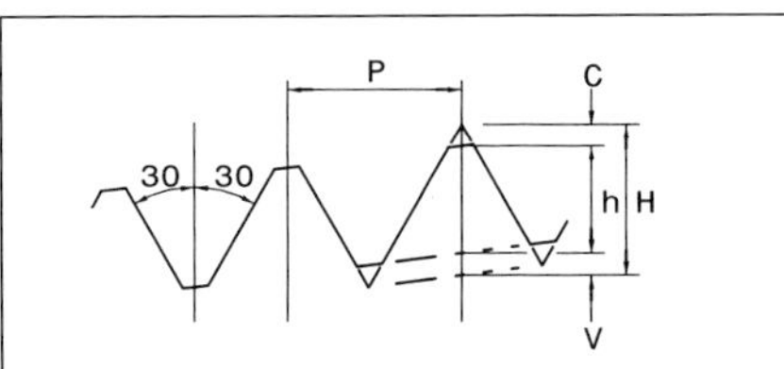

27TPI = 0.096P
18TPI = 0.088P
14TOPI   = 0.078P
11.5TPI = 0.073P
8TPI   = 0.062P
H   peak-to-peak height
        =0.866P
h   height of thread
        =0.8P
V   as for C
Taper   1 in 16 on diameter

ACME THREADS, IMPERIAL

CN   crest, nut
CS   crest, screw

Acme threads: nominal outline no axial clearance

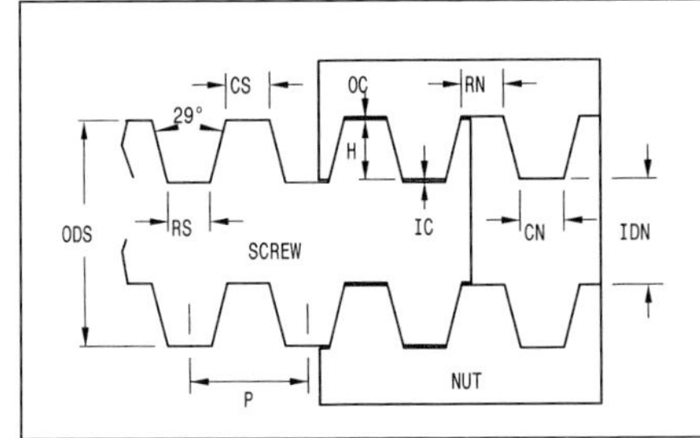

FAA   fit allowance, acme
H   basic depth
IC   inside clearance
IDN   inside diameter, nut
OC   outside clearance
ODS   outside diameter, screw
P   pitch
RN   root, nut
RS   root, screw
TPI   threads per inch

Table 9.1 Acme threads: dimensions

ODS	TPI	IDN	CN	OC	IC	FAA
0.250	16	0.187	0.023	0.005	0.005	0.0010
0.312	14	0.241	0.026	0.005	0.005	0.0010
0.375	10	0.275	0.037	0.010	0.010	0.0013
0.437	10	0.337	0.037	0.010	0.010	0.0013
0.500	10	0.400	0.037	0.010	0.010	0.0015
0.625	8	0.500	0.046	0.010	0.010	0.0016
0.750	6	0.583	0.062	0.010	0.010	0.0018
0.875	6	0.708	0.062	0.010	0.010	0.0019
1.000	6	0.833	0.062	0.010	0.010	0.0021

Abbreviations: see drawing above.
Dimensions are in inches
See formulas for determining other values

Formulas

Nut:

This is made to fixed dimensions, enabling a common tap to be used independent of the class of fit required.

$$CN = 0.3707P$$
$$RN = 0.3707P - 0.517 \times OC$$
$$IDN = ODS - 2H$$
$$\text{Depth of thread} = 0.5P + OC$$

Screw:

As the screw is likely to be screw-cut on the lathe, class of fit is controlled by varying the dimensions of the screw. The value given for the FAA is for a free-running fit. Using suitable equipment, this can be reduced by up to 50% or more in order to limit axial clearance (backlash).

$$CS = 0.3707P - 2FAA$$
$$RS = 0.3707P - 0.517 \times IC + 2FAA$$
$$H = 0.5P$$
$$\text{Depth of thread} = 0.5P + IC$$
$$\text{Axial clearance} = 2FAA$$

Note that H is the depth of engagement.

ISO METRIC TRAPEZOIDAL THREADS

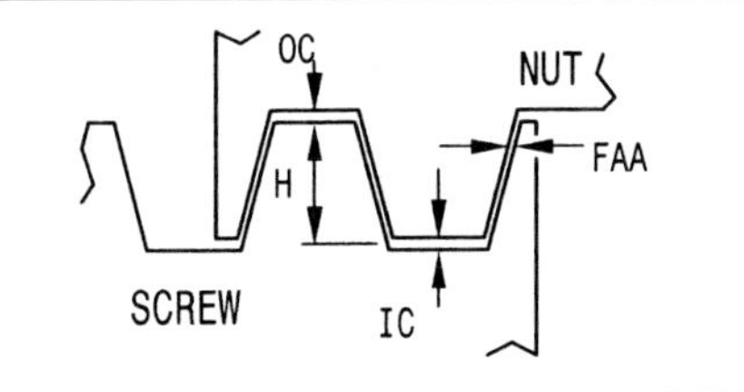

Acme thread fit allowance (FAA)

In the metric system the comparable thread to the imperial acme thread is the ISO metric trapezoidal thread. This closely resembles the acme thread so the references and associated drawing can be assumed to apply to the acme thread as well.

Differences

1. The internal angle is increased from 29° to 30°.

2. To vary the fit, the same tool is used to cut the screw thread, but its depth is increased to increase the width of the cut. This differs from the acme thread, where a wider tool is used at the same depth. Dimension IC is therefore varied by an amount (the trapezoidal fit

allowance, or FAT), as is the depth of cut, to control fit. The root width of the screw (RS) is therefore a constant, but the crest of the screw (CS) varies.

Table 9.2. ISO metric trapezoidal thread: dimensions

ODS	P	IDN	CN	OC	IC
8	1.5	6.5	0.55	0.15	0.15
10	1.5	8.5	0.55	0.15	0.15
10	2	8.0	0.73	0.25	0.25
12	2	10.0	0.73	0.25	0.25
12	3	9.0	1.10	0.25	0.25
16	2	14.0	0.73	0.25	0.25
16	4	12.0	1.46	0.25	0.25
20	2	18.0	0.73	0.25	0.25
20	4	16.0	1.46	0.25	0.25
24	3	21.0	1.10	0.25	0.25
24	5	19.0	1.83	0.25	0.25

Abbreviations: as above.
Determining other values: see formulas below.
Dimensions: in millimeters.

A value of 0.10mm for the FAT will give an axial clearance of 0.05mm. Other values of FAT will affect the axial clearance by the same proportion, i.e. 50%.

Formulas

Nut:

$CN = 0.366P$

$RN = 0.366P - 0.536 \times OC$

$IDN = ODS - 2H$

$Depth\ of\ thread = 0.5P + OC$

Screw:

$CS = 0.366P - 0.536 \times FAT$

$RS = 0.366P - 0.536 \times IC$

$H = 0.5P$

$Depth\ of\ thread = 0.5P + IC + FAT$

$Axial\ clearance = 0.536 \times FAT$

Note that H is the depth of engagement.

ISO metric trapezoidal thread fit allowance (FAT)

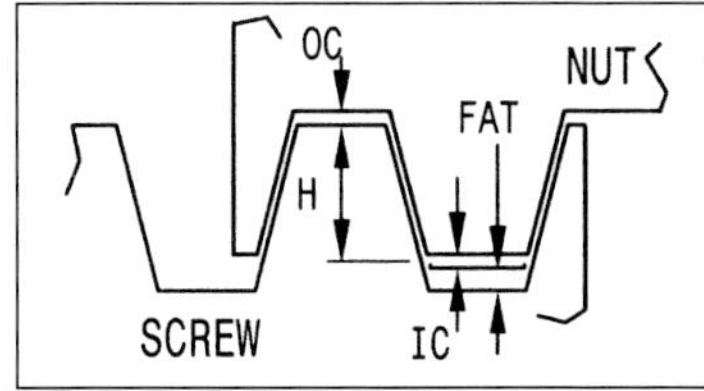

Abbreviations: as above

ROOT AND CREST RADII

For ease of manufacture, a radius in each corner of the root (screw and/or nut) is permissible up to the value of the inside clearance (IC) or outside clearance (OC) for both acme and trapezoidal screws. The crest can also have a radius of up to half this value.

CHAPTER 10
SCREW CUTTING ON THE LATHE

CHANGEWHEEL COMBINATIONS

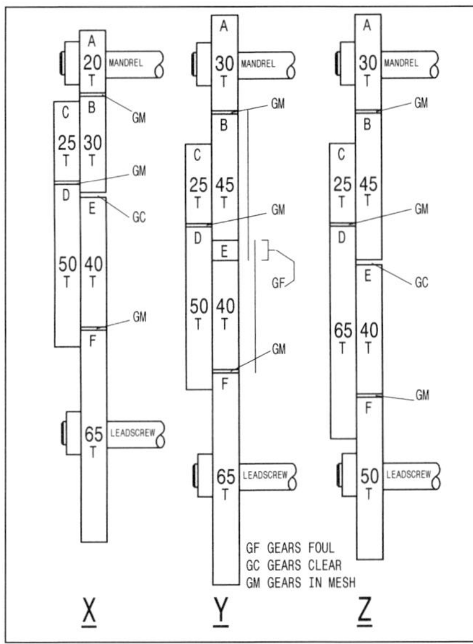

10-01: Changewheel combinations

NOTES

The combinations given in the tables below have been proven to be mathematically correct but their acceptability will depend on the machine in question. The most obvious problem is having insufficient changewheels, either because they were not initially supplied or because they have been lost. If the gears listed are too large, this may result in a center distance being too large, making it impossible to fit them to the quadrant.

These combinations are in no way exclusive because others can be used to arrive at the same result, using other available gears. Even so, combinations that are mathematically correct may present problems in fitting.

Consider example X in the drawing, which achieves 39TPI on a lathe with an 8TPI leadscrew. If, say, a 20-tooth gear is not available, the 1:1.5 ratio (20:30) could also be achieved by using the 30-tooth gear as a driver and driving a 45-tooth gear (30:45).

However, example Y shows that gear B will then foul gear E, rather than providing the essential clearance. This can be proved mathematically as the total number of teeth on gears B and E (45 + 40 = 85) is greater than the sum of the teeth on gears C and D (25 + 50 = 75).

The lack of a 20-tooth gear does not mean that a 39TPI screw cannot be cut, because the three drivers and three driven can be used in any order and any pairing. Consider therefore, interchanging the 50-tooth driven gear with the 65-tooth driven gear , as in example Z. This shows that the sum of the teeth on gears C and D, being 90, is greater than that of B and E, which total 85, thus ensuring that there is the required clearance between B and E.

From this example, it can be seen that more than one combination of gears meet a given requirement, and that the drivers and driven can be use in any order and in any pairing, provided that the order is mechanically acceptable. For most threads, more than one combination will provide the same result, and there are often many more.

Therefore, if a quoted gear is not available, look for other combinations that will provide the same result. For example, in example X, the 1:2 ratio (25:50) could be replaced by 35- and 70-tooth gears. Using a standard gear set ,there are in fact 19 combinations that will achieve 39TPI.

FOUR TPI LEADSCREW

Although larger lathes may have a 4TPI leadscrew, this is rarely a factor in the small workshop, so no table for this has been included. However, there is a simple method of arriving at the required result. If you take a combination from the table for an 8TPI leadscrew and set it up on a lathe with a 4TPI leadscrew the resulting thread will provide a TPI value of half that in the 8TPI listing. In this way, a 40TPI combination will result in a 20TPI thread being cut.

As the table goes up to 60TPI, this will give values up to 30TPI. For values above this, take the required value and either halve the size of one of the driver gears or double the size of one of the driven gears. However, this may be more difficult than it appears because driven gears may need to be larger than those available and driver gears may need to be too small, In addition, in some cases, duplicate gears may be required. In this case an alternative set-up would have to be found, probably a three driver/three driven set-up in place of a two driver/two driven set-up. In this case, any pair of gears with a 1:2 ratio, used in place of the idler, will provide the required adjustment.

GEAR SIZES IN INCREMENTS OF FOUR TEETH

If your lathe has gears in increments of four teeth rather than five, the tables can still be easily used. Take each quoted size and reduce the value by ⅘ (e.g. for a gear chain of 25, 35, 40 and 65, the reduced values would be 20, 28, 32 and 52). This retains the ratios as listed and in no way alters the TPI to be cut.

FORMULA

If the occasion arises when you have to determine your own gear set-up, the following formula will enable you to check your chosen values:

$$CTPI = LTPI \times \frac{Dn1 \times Dn2 \times Dn3}{Dr1 \times Dr2 \times Dr3}$$

where:
CTPI = TPI to be cut
LTPI = leadscrew TPI
Dn1, Dn2, Dn3 = driven gears
Dr1, Dr2, Dr3 = driver gears

Omit Dn3 and Dr3 for a two-driver/two-driven set-up and, for a single driver/single drive set-up, omit Dn3 and DR3 as well.

SCREW-CUTTING CHANGEWHEEL COMBINATIONS

The tables that follow give typical changewheel combinations for cutting imperial threads on lathes having 6TPI, 8TPI or 10TPI leadscrews. Also, a further table gives changewheel sizes for cutting metric threads on a lathe having an 8TPI leadscrew. In this case though, as calculating alternative combinations is a very complex task, more than one combinations is given to assist those who do not have a full set of changewheels available.

Table 10.1. For leadscrew with 10TPI

TPI	Mandrel	First stud		Second stud		
		Driven	Driver	Driven	Driver	Leadscrew
2	50	30	60	idler		20
3	50	30	40	idler		20
4	50	idler		idler		20
5	50	idler		idler		25
6	50	idler		idler		30
7	50	idler		idler		35
8	50	idler		idler		40
9	50	idler		idler		45
10	50	20	30	idler		75
11	50	idler		idler		55
12	25	idler		idler		30
13	20	65	50	idler		20
14	25	idler		idler		35
15	20	idler		idler		30
16	25	idler		idler		40
17	50	idler		idler		85
18	25	idler		idler		45
19	20	idler		idler		38
20	20	idler		idler		40
21	20	70	50	idler		30
22	25	idler		idler		55
24	20	60	50	idler		40
25	20	idler		idler		50
26	25	idler		idler		65
27	20	30	25	idler		45
28	20	70	50	idler		40
30	20	30	25	idler		50
32	25	40	30	idler		60
33	20	30	25	idler		55
34	25	idler		idler		85
35	20	35	25	idler		50
36	20	30	25	idler		60
38	20	38	30	idler		60
39	20	30	25	idler		65
40	20	40	25	idler		50
42	25	35	20	idler		60
44	20	40	25	idler		55

TPI	Mandrel	First stud		Second stud		
		Driven	Driver	Driven	Driver	Leadscrew
45	25	45	20	idler		50
48	20	40	25	idler		60
50	20	50	30	idler		60
56	20	40	25	idler		70
60	20	50	25	idler		60

Table 10.2. For leadscrew with 8TPI

TPI	Mandrel	First stud		Second stud		
		Driven	Driver	Driven	Driver	Leadscrew
2	50	25	60	idler		30
3	50	25	60	idler		45
4	60	idler		idler		30
5	60	30	40	idler		50
6	60	idler		idler		45
7	40	idler		idle		35
8	60	40	50	idler		75
9	40	idler		idler		45
10	40	idler		idler		50
11	40	idler		idler		55
12	40	idler		idler		60
13	40	idler		idler		65
14	20	idler		idler		35
15	40	45	30	idler		50
16	20	idler		idler		40
17	40	idler		idler		85
18	20	idler		idler		45
19	20	38	40	idler		50
20	20	idler		idler		50
21	20	70	40	idler		30
22	20	idler		idler		55
24	20	idler		idler		60
25	20	25	20	idler		50
26	20	idler		idler		65
27	20	30	20	idler		45
28	20	35	20	idler		40
30	20	50	40	idler		60
32	20	40	25	idler		50
33	20	45	30	idler		55

TPI	Mandrel	First stud		Second stud		
		Driven	Driver	Driven	Driver	Leadscrew
34	20	idler		idler		85
35	20	50	40	idler		70
36	20	40	20	idler		45
38	20	38	40	50	30	60
39	20	30	25	50	40	65
40	20	40	20	idler		50
42	20	60	40	idler		70
44	20	55	30	idler		60
45	20	45	40	50	30	60
48	20	50	25	idler		60
50	20	50	30	idler		75
56	20	50	25	idler		70
60	20	50	20	idler		60

Table 10.3. For leadscrew with 6TPI

TPI	Mandrel	First stud		Second stud		
		Driven	Driver	Driven	Driver	Leadscrew
2	60	idler		idler		20
3	60	idler		idler		30
4	60	idler		idler		40
5	60	idler		idler		50
6	60	30	20	idler		40
7	30	idler		idler		35
8	30	idler		idler		40
9	30	idler		idler		45
10	30	idler		idler		50
11	30	idler		idler		55
12	30	idler		idler		60
13	30	idler		dler		65
14	30	idler		idler		70
15	30	50	40	idler		60
16	30	40	35	idler		70
17	30	idler		idler		85
18	30	60	50	idler		75
19	20	38	30	idler		50
20	20	40	30	idler		50
21	40	60	30	idler		70
22	20	40	30	idler		55

TPI	Mandrel	First stud		Second stud		
		Driven	Driver	Driven	Driver	Leadscrew
24	20	40	30	idler		60
25	20	50	45	idler		75
26	20	40	30	idler		65
27	20	75	60	80	50	45
28	20	40	30	idler		70
30	20	50	30	idler		60
32	20	40	30	idler		80
33	20	60	30	idler		55
34	20	40	30	idler		85
35	20	50	30	idler		70
36	20	60	25	idler		50
38	20	38	40	50	30	80
39	30	60	35	70	40	65
40	20	50	45	60	35	70
42	20	60	30	idler		70
44	20	40	25	50	30	55
45	20	50	40	60	35	70
48	20	60	30	idler		80
50	20	50	40	80	45	75
56	20	70	30	idler		80
60	20	50	30	60	35	70

FOR LEADSCREW WITH 8TPI

EASY SET-UP ARRANGEMENT

This arrangement provides all the listed TPI values without having to move either the first or second stud. Only the quadrant needs swinging to engage the gear on the mandrel with that on the first stud.

For the 19TPI setting the 38-tooth gear meshes loosely with the 55-tooth gear on the second stud but is adequate for the task.

The system requires 2 × 40-tooth gears for some values and the 25/12 tooth fine-feed cluster for 40TPI.

The fine feed equates to 0.0022in feed per revolution.

COMPLEX PITCHES

The need to cut threads with complex values usually results from the need to cut metric threads on an imperial lathe, or vice versa. Typically a thread of 1.5mm pitch would require an imperial thread of 16.93333 TPI. Worms require similarly complex values.

When cutting complex pitches it is normally accepted that the exact value will not be possible but that sufficiently close values will be achievable. The number of

Table 10.4. Easy set-up arrangement for leadscrew with 8TPI

TPI	Mandrel	First stud		Second stud	Lead screw
		Driven	Driver	Idler	
8	40	40	20	75	20
9	40	45	20	75	20
10	40	50	20	75	20
11	40	55	20	75	20
12	40	60	20	75	20
14	40	70	20	75	20
16	30	60	40	55	40
18	20	45	40	55	40
19	20	50	40	55	38
20	20	50	40	55	40
24	20	60	40	55	40
26	20	65	40	55	40
28	20	70	40	55	40
30	20	75	40	55	40
40	12	60	40	55	40
Fine feed				DR/DN	
455	12	65	25	70 & 20	75

combinations giving possible values can therefore be large.

The combinations that follow are only a few of the closest possible and have been taken from a complete list worked out by a dedicated computer program. Many require a three-driver/three-driven combination, but reasonably close values are also possible in some cases with a two-driver/two-driven combination. The listing is based on 20- to 75-tooth wheels, in steps of 5 and 38, a common standard set. A 21-tooth gear is also included as this is a frequently available extra specifically for cutting metric values. The computer program proves that, for many values, equal or better results can be achieved without the use of the 21-tooth wheel.

NOTES REGARDING TABLE

1. The dimensional error in Table 10.5 is quoted for 10 pitches, as in most cases this will be comparable to the amount of engagement between the internal and external threads. The percentage error is also given.

2. The table includes the closest possible values, both in the plus and minus directions. In most cases there will be more than one combination that provides these values, but for most pitches only one is included.

3. The closest values frequently require a 38-tooth or the less standard 21-tooth gear. Because of this, combinations that do not use these have also been

included, although they are less accurate.

4. Two-driver/two-driven combinations are usually less accurate, but some are given where the error is reasonable because they are easier to set-up and require fewer gears (a benefit to those with incomplete gear sets).

5. While it would be good practice to aim for the most accurate combination, any of those given should be adequate for all but the most demanding applications.

6. No sizes are given for idlers as they are not important, having no effect on the resulting pitch. Choose a size that best suits the space available.

7. The notes at the beginning of this section also largely apply to the combinations published here.

The above notes, except that regarding the use of 38- and 21-tooth gears, apply equally to the changewheel combinations given for DP and Module worms in Chapter 11.

METRIC CHANGEWHEEL COMBINATIONS

The Module gear system is based on metric dimensions. However, as the tables are for worms being cut on an imperial lathe, the imperial conversion for circular pitch is listed. The circular pitch in metric terms can be calculated from the following formula:

$$CP = \pi \times Mod$$

where:

$\pi = 3.1416$.

Table 10.5. Metric screw-cutting changewheel combinations for a lathe with an 8TPI imperial leadscrew

Pitch				Mandrel	First stud		Second stud		Leadscrew
Aim	Result	Error ×10 mm	% Error	Driver	Driven	Driver	Driven	Driver	Driven
0.50	0.49956	−0.00437	−0.087	45	55	30	60	25	65
0.50	0.49989	−0.00112	−0.022	20	38	35	45	25	65
0.50	0.50006	0.00062	0.012	21	50	45	40	20	60
0.50	0.50030	0.00303	0.061	65	55	35	70	20	75
0.60	0.59987	−0.00135	−0.022	35	38	30	45	20	65
0.60	0.59997	−0.00031	−0.005	20	45	38	55	40	65
0.60	0.60008	0.00075	0.013	21	50	45	40	30	75
0.60	0.60036	0.00364	0.061	65	50	30	55	20	75
0.70	0.69980	−0.00204	−0.029	45	35	30	50	20	70
0.70	0.70009	0.00088	0.012	35	40	45	50	21	75
0.70	0.70025	0.00251	0.036	40	38	55	70	20	75
0.70	0.70042	0.00424	0.061	65	50	35	55	20	75

Pitch				Mandrel	First stud		Second stud		Leadscrew
Aim	Result	Error ×10 mm	% Error	Driver	Driven	Driver	Driven	Driver	Driven
0.75	0.74996	−0.00039	−0.005	38	55	50	45	20	65
0.75	0.74996	−0.00039	−0.005	25	45	40	55	38	65
0.75	0.75009	0.00094	0.012	45	40	35	50	21	70
0.75	0.75045	0.00455	0.061	65	55	35	70	30	75
0.80	0.79957	−0.00435	−0.054	25	60	55	65	50	70
0.80	0.79982	−0.00180	−0.022	40	38	35	45	20	65
0.80	0.80010	0.00100	0.013	20	25	21	40	30	50
0.80	0.80010	0.00100	0.013	21	50	45	idler		75
0.90	0.89994	−0.00057	−0.006	25	21	20	60	50	70
0.90	0.89995	−0.00047	−0.005	38	30	40	55	20	65
0.90	0.89995	−0.00047	−0.005	60	45	38	55	20	65
0.90	0.90084	0.00843	0.094	65	60	55	70	25	75
1.00	0.99978	−0.00225	−0.022	35	45	50	38	20	65
1.00	1.00012	0.00125	0.012	21	40	30	idler		50
1.00	1.00012	0.00125	0.012	45	40	21	idler		75
1.00	1.00012	0.00125	0.012	30	40	45	50	35	75
1.10	1.09957	−0.00433	−0.039	40	55	50	35	25	75
1.10	1.09994	−0.00057	−0.005	38	45	40	30	20	65
1.10	1.10003	0.00025	0.002	35	55	70	45	21	60
1.10	1.10014	0.00138	0.013	45	50	55	40	21	75
1.20	1.19935	−0.00652	−0.054	55	40	50	65	25	70
1.20	1.19973	−0.00270	−0.022	40	38	35	45	30	65
1.20	1.19992	−0.00076	−0.006	50	21	20	45	25	70
1.20	1.20015	0.00150	0.013	45	25	30	50	21	60
1.25	1.24972	−0.00281	−0.022	50	38	35	45	25	65
1.25	1.25006	0.00063	0.005	30	35	38	65	55	70
1.25	1.25016	0.00156	0.013	45	40	21	idler		60
1.25	1.25016	0.00156	0.013	45	40	35	50	30	60
1.30	1.29929	−0.00707	−0.054	50	40	55	60	25	70
1.30	1.29971	−0.00292	−0.022	20	38	35	idler		45
1.30	1.29993	−0.00067	−0.005	40	30	38	45	20	55

Pitch				Mandrel	First stud		Second stud		Leadscrew
Aim	Result	Error ×10 mm	% Error	Driver	Driven	Driver	Driven	Driver	Driven
1.30	1.30016	0.00163	0.013	65	40	45	50	21	75
1.40	1.39959	−0.00408	−0.029	60	35	45	50	20	70
1.40	1.39969	−0.00315	−0.022	70	38	35	45	20	65
1.40	1.39991	−0.00088	−0.006	50	21	20	45	25	60
1.40	1.40017	0.00175	0.012	45	25	35	50	21	60
1.50	1.49966	−0.00337	−0.022	35	30	40	38	25	65
1.50	1.49992	−0.00078	−0.005	38	45	50	55	40	65
1.50	1.50019	0.00188	0.012	21	50	45	idler		40
1.50	1.50091	0.00909	0.061	65	55	30	idler		75
1.60	1.59964	−0.00360	−0.022	70	38	40	45	20	65
1.60	1.60020	0.00200	0.013	45	25	35	50	30	75
1.60	1.60020	0.00200	0.013	30	25	21	idler		50
1.60	1.60020	0.00200	0.013	45	25	21	idler		75
1.75	1.74949	−0.00510	−0.029	45	35	30	idler		70
1.75	1.74961	−0.00394	−0.022	70	38	35	45	25	65
1.75	1.75009	0.00088	0.005	55	25	30	65	38	70
1.75	1.75022	0.00219	0.012	45	20	21	50	35	60
1.80	1.79960	−0.00405	−0.022	35	38	40	idler		65
1.80	1.79991	−0.00093	−0.005	40	55	60	45	38	65
1.80	1.80022	0.00225	0.013	30	40	45	25	21	50
1.80	1.80064	0.00637	0.035	70	40	35	45	25	60
2.00	1.99955	−0.00450	−0.022	70	38	50	45	20	65
2.00	2.00025	0.00250	0.012	70	40	45	50	30	75
2.00	2.00025	0.00250	0.012	35	40	45	25	20	50
2.00	2.00025	0.00250	0.012	60	40	21	idler		50
2.25	2.24949	−0.00506	−0.022	70	38	50	40	20	65
2.25	2.24988	−0.00117	−0.005	50	30	38	55	40	65
2.25	2.25028	0.00281	0.012	45	20	21	40	30	50
2.25	2.25136	0.01364	0.061	65	55	45	idler		75

Pitch				Mandrel	First stud		Second stud		Leadscrew
Aim	Result	Error ×10 mm	% Error	Driver	Driven	Driver	Driven	Driver	Driven
2.50	2.49944	−0.00562	−0.022	70	38	50	45	25	65
2.50	2.50013	0.00126	0.005	55	35	60	65	38	70
2.50	2.50031	0.00313	0.013	45	30	21	idler		40
2.50	2.50031	0.00313	0.013	35	40	45	idler		50
2.75	2.74892	−0.01082	−0.039	75	35	50	45	20	55
2.75	2.74892	−0.01082	−0.039	50	35	60	45	25	55
2.75	2.74986	−0.00142	−0.005	50	30	38	45	40	65
2.75	2.75034	0.00344	0.012	55	20	30	40	21	50
3.00	2.99933	−0.00675	−0.022	70	45	50	38	30	65
3.00	2.99933	−0.00675	−0.022	50	30	35	38	40	65
3.00	3.00038	0.00375	0.012	45	25	21	idler		40
3.00	3.00038	0.00375	0.012	45	25	30	40	35	50
3.50	3.49898	−0.01020	−0.029	60	35	45	idler		70
3.50	3.49921	−0.00787	−0.022	70	38	50	45	35	65
3.50	3.50018	0.00176	0.005	55	25	38	35	30	65
3.50	3.50044	0.00437	0.012	60	20	35	40	21	50

THREAD INDICATOR

CUTTING THREADS

Most lathe-users will at some time use the lathe to cut a thread. Once this has been done a few times the task is likely to be understood. However, even at this stage, reference to data regarding the use of the thread indicator is likely to be needed.

HALF NUT LIMITATIONS

It is frequently implied that the purpose of the half nut is to aid thread cutting. This is not so. Its purpose is to enable the saddle to be rapidly moved along the lathe's bed unrestricted by the leadscrew. However, it is true that its use only becomes critical when cutting screw threads, which is when the need for the thread indicator arises.

WHEN TO CLOSE THE HALF NUT

Except when the thread being cut is a multiple of the leadscrew TPI (e.g. 8, 16, 24, etc.), the cutter is unlikely to line up with the already partly cut thread if the half nut is closed randomly, so the following must be worked to:

- For other threads that are a multiple of 4 (4, 12, 20, etc.), the half nut can be closed at every full division or at half divisions if they are included on the dial.

- For all other even numbers, the half nut can be closed at any numbered position.
- For odd numbers, the half nut can be closed at any two opposite marks, 1 and 3, or 2 and 4.
- For half number, such as 9½ TPI, the half nut should always be closed at the same position. For more complex values the half nut should not be disengaged.

OTHER LEADSCREW PITCHES

For other leadscrew pitch and indicator gear combinations, adopt the following approach if the lathe-manufacturer's data is not available.

Determine the amount of saddle movement for one revolution of the indicator dial. To do this, divide the number of gear teeth by the TPI of the leadscrew. For example, an 18-tooth gear with a leadscrew of 6TPI will result in a 3in movement for one revolution of the dial.

Calculate the number of thread pitches, both for the leadscrew and the thread being cut over the calculated saddle movement. For example, for a 6TPI leadscrew and a 14TPI cut thread, the values will be 18 threads and 42 threads over 3in. Divide each value by the number of divisions on the dial, e.g. for four divisions this will give 4.5 and 10.5 threads. Calculate the number of threads, both of the leadscrew and the thread being cut, at each marking, as follows for the example:

Mark	1	2	3	4	1
Leadscrew	0	4.5	9.0	13.5	18.0
Cut thread	0	10.5	21.0	31.5	42.0

The half nut can be closed at any position where both values are whole numbers, i.e. positions 1 and 3 in this example.

CHAPTER 11
WORM CUTTING

The requirements for the tool to cut the worm on the lathe are quite exacting.

INTERNAL ANGLE

The tool must be ground to have an internal angle of twice the pressure angle of the mating wormwheel, e.g. 40° and 29° for the common pressure angles of 20° and 14.5°.

TOOL-TIP WIDTH

The tool is not ground to a point but has a flat, the width of which is given in the tables below. However, as the bottom of the groove only provides clearance over the wormwheel outside diameter, its depth is not critical. For this reason, a tool for a smaller pitch, with a narrower end, could be used for a larger pitch by cutting the groove deeper. In this case, the depth would need to be calculated as that given in the tables would no longer apply.

HELIX ANGLE

Since the diameter of the worm in no way affects the resulting ratio with the wormwheel, it can be made any diameter to suit the space available. As the helix angle is diameter dependent it is therefore variable, and the tool must be made to suit the angle. It is easier to grind the shape on to a round tool bit mounted into a holder, so that it can be rotated within the holder to the helix angle required. The angle (Ha) is calculated using the following formula.

$$\tan Ha = \frac{\text{Pitch}}{\pi \times \text{PCD}}$$

where:

$\pi = 3.1416$

PCD = pitch circle diameter.

Note that it is the pitch circle diameter, not the outside diameter, that is used in the formula.

A reference flat on the side of the toolbit (as shown in the drawing) will help in setting it to the angle required.

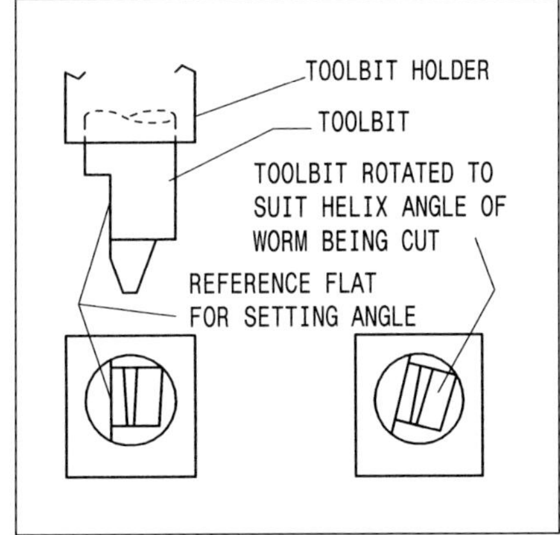

WHOLE DEPTH VALUE

The values listed for diametral pitch (DP) and metric module (MOD) worms are based on a whole depth value (H) for the worm, as follows:

DP worms:

$$H = \frac{2.4}{DP}$$

MOD worms:

$$H = 2.4 \times MOD$$

Some published data may use an early standard of 2.16, rather than 2.4, for the 'whole depth ratio'. This is not normally stated but can be checked by multiplying the value given for H by the DP to see if this gives a value of 2.16 or another value. This gives less clearance between the outside diameter of the wormwheel and the base of the groove in the worm, and also affects the size of the flat on the end of the toolbit.

PITCH CIRCLE/OUTSIDE DIAMETERS

The pitch circle diameter can be calculated by taking the outside diameter and deducting from this twice the value of the addendum given in the table. Alternatively, if the pitch circle diameter is chosen first, add twice the addendum to this to arrive at the outside diameter.

CENTER DISTANCE

The center distance between worm and wormwheel can be calculated using the formula for center distance in Chapter 12 (see page 77). The spindles are, of course, at right angles and not parallel.

CIRCULAR PITCH/THREADS PER INCH

As is normal for worms, the following tables quote the pitch in terms of circular pitch, to correspond with the value for the wormwheel. However, for the changewheel tables that follow, the pitch is quoted in terms of threads per inch as this is more appropriate to setting up changewheels for cutting a worm on the lathe.

The formulas for circular pitch in terms of threads per inch (TPI) are:

DP worms:

$$TPI = \frac{DP}{\pi}$$

MOD worms:

$$TPI = \frac{25.4}{\pi \times MOD}$$

TOOL TIP SIZE

The tool tip size for other pressure angles and or whole depth ratios can be calculated from the following formulas.

Tool tip width (TTW) is calculated as follows.

DP worms:

$$TTW = \frac{\pi - 4(WDR - 1)\tan PA}{2DP}$$

MOD worms:

$$TTW = MOD \left(\frac{\pi}{2DP} - 2(WDR - 1)\tan PA \right)$$

where:

PA = pressure angle

WDR = whole depth ratio

See also Chapter 12 for other relevant formula and definitions.

Table 11.1. DP worms: 20° pressure angle, 2.4 height ratio

DP	Tool tip width	Depth of cut	Worm tip width	Addendum	Circular pitch
12	0.046	0.200	0.070	0.083	0.262
14	0.039	0.171	0.060	0.071	0.224
16	0.034	0.150	0.053	0.063	0.196
18	0.031	0.133	0.047	0.056	0.175
20	0.028	0.120	0.042	0.050	0.157
22	0.025	0.109	0.038	0.045	0.143
24	0.023	0.100	0.035	0.042	0.131
28	0.020	0.086	0.030	0.036	0.112
32	0.017	0.075	0.026	0.031	0.098
36	0.015	0.067	0.023	0.028	0.087
40	0.014	0.060	0.021	0.025	0.079
44	0.013	0.055	0.019	0.023	0.071
48	0.011	0.050	0.018	0.021	0.065
64	0.009	0.038	0.013	0.016	0.049

Dimensions in inches

Table 11.2. DP worms: 14.5° pressure angle, 2.4 height ratio

DP	Tool tip width	Depth of cut	Worm tip width	Addendum	Circular pitch
12	0.071	0.200	0.088	0.083	0.262
14	0.060	0.171	0.075	0.071	0.224
16	0.053	0.150	0.066	0.063	0.196
18	0.047	0.133	0.059	0.056	0.175
20	0.042	0.120	0.053	0.050	0.157
22	0.038	0.109	0.048	0.045	0.143
24	0.035	0.100	0.044	0.042	0.131
28	0.030	0.086	0.038	0.036	0.112
32	0.026	0.075	0.033	0.031	0.098
36	0.024	0.067	0.029	0.028	0.087
40	0.021	0.060	0.026	0.025	0.079
44	0.019	0.055	0.024	0.023	0.071
48	0.018	0.050	0.022	0.021	0.065
64	0.013	0.038	0.016	0.016	0.049

Dimensions in inches

Table 11.3. MOD worms: 20° pressure angle, 2.4 height ratio

DP	Tool tip width	Depth of cut	Worm tip width	Addendum	Circular pitch
0.40	0.22	0.96	0.34	0.40	1.26
0.50	0.28	1.20	0.42	0.50	1.57
0.60	0.33	1.44	0.51	0.60	1.88
0.70	0.39	1.68	0.59	0.70	2.20
0.75	0.41	1.80	0.63	0.75	2.36
0.80	0.44	1.92	0.67	0.80	2.51
0.90	0.50	2.16	0.76	0.90	2.83
1.00	0.55	2.40	0.84	1.00	3.14
1.25	0.69	3.00	1.05	1.25	3.93
1.50	0.83	3.60	1.26	1.50	4.71
1.75	0.97	4.20	1.47	1.75	5.50
2.00	1.10	4.80	1.69	2.00	6.28

Dimensions in millimeters

Table 11.4. MOD worms: 14.5° pressure angle, 2.4 height ratio

DP	Tool tip width	Depth of cut	Worm tip width	Addendum	Circular pitch
0.40	0.34	0.96	0.42	0.40	1.26
0.50	0.42	1.20	0.53	0.50	1.57
0.60	0.51	1.44	0.63	0.60	1.88
0.70	0.59	1.68	0.74	0.70	2.20
0.75	0.64	1.80	0.79	0.75	2.36
0.80	0.68	1.92	0.84	0.80	2.51
0.90	0.76	2.16	0.95	0.90	2.83
1.00	0.85	2.40	1.05	1.00	3.14
1.25	1.06	3.00	1.32	1.25	3.93
1.50	1.27	3.60	1.58	1.50	4.71
1.75	1.48	4.20	1.84	1.75	5.50
2.00	1.69	4.80	2.11	2.00	6.28

Dimensions in millimeters

Table 11.5. DP worms: screw-cutting changewheel combinations for a lathe with an 8TPI imperial leadscrew

	Worm			Mandrel	1st stud		2nd stud		Lead screw
DP	Required TPI	Resulting TPI	% error	Driver	Driven	Driver	Driven	Driver	Driven
12	3.81972	3.81818	−0.040	55	30	40	idler		35
12	3.81972	3.81949	−0.006	65	35	75	38	40	70
12	3.81972	3.82041	0.018	70	20	50	45	35	65
12	3.81972	3.82041	0.018	70	30	75	45	35	65
14	4.45634	4.45558	−0.017	45	20	38	25	21	40
14	4.45634	4.45714	0.018	70	20	40	30	25	65
14	4.45634	4.45714	0.018	70	30	50	idler		65
14	4.45634	4.45714	0.018	70	45	75	idler		65
16	5.09296	5.09091	−0.040	55	idler		idler		35
16	5.09296	5.09259	−0.007	60	25	45	50	40	55
16	5.09296	5.09265	−0.006	65	35	50	38	45	70
16	5.09296	5.09388	0.018	70	20	35	30	25	65
18	5.72958	5.72727	−0.040	55	35	40	idler		45
18	5.72958	5.72923	−0.006	65	35	50	38	40	70
18	5.72958	5.73057	0.017	65	20	45	40	21	55
18	5.72958	5.73061	0.018	70	30	50	45	35	65
20	6.36620	6.36364	−0.040	55	35	60	idler		75
20	6.36620	6.36581	−0.006	65	35	45	38	40	70
20	6.36620	6.36632	0.002	50	21	38	30	25	60
20	6.36620	6.36735	0.018	35	65	70	idler		30
22	7.00282	7.00000	−0.040	40	idler		idler		35
22	7.00282	7.00337	0.008	55	25	45	40	30	65
22	7.00282	7.00337	0.008	55	20	45	50	30	65
22	7.00282	7.00408	0.018	50	55	70	30	35	65
24	7.63944	7.63636	−0.040	55	35	40	idler		60
24	7.63944	7.63889	−0.007	40	50	60	25	30	55
24	7.63944	7.63897	−0.006	65	35	50	38	30	70
24	7.63944	7.64082	0.018	70	20	35	45	25	65
28	8.91268	8.90909	−0.040	55	35	40	idler		70
28	8.91268	8.91117	−0.017	45	20	38	40	21	50
28	8.91268	8.91429	0.018	70	30	50	40	20	65
28	8.91268	8.91429	0.018	25	65	70	idler		30
32	10.18592	10.18182	−0.040	55	idler		idler		70
32	10.18592	10.18519	−0.007	60	25	45	55	30	75
32	10.18592	10.18530	−0.006	65	35	45	38	25	70
32	10.18592	10.18776	0.018	35	60	70	20	25	65
36	11.45916	11.45833	−0.007	60	25	40	55	30	75
36	11.45916	11.45846	−0.006	65	35	40	38	25	70
36	11.45916	11.46122	0.018	70	30	35	45	25	65

	Worm			Mandrel	1st stud		2nd stud		Lead screw
DP	Required TPI	Resulting TPI	% error	Driver	Driven	Driver	Driven	Driver	Driven
36	11.45916	11.46199	0.025	38	35	45	idler		70
40	12.73240	12.73162	−0.006	65	35	45	38	20	70
40	12.73240	12.73469	0.018	70	30	35	40	20	65
40	12.73240	12.73469	0.018	35	60	70	idler		65
40	12.73240	12.73469	0.018	70	20	35	65	25	75
44	14.00564	14.00000	−0.040	40	idler		idler		70
44	14.00564	14.00673	0.008	55	20	45	65	21	70
44	14.00564	14.00673	0.008	55	40	45	50	30	65
44	14.00564	14.00816	0.018	70	30	35	55	25	65
48	15.27887	15.27273	−0.040	55	35	25	idler		75
48	15.27887	15.27778	−0.007	45	25	40	55	30	75
48	15.27887	15.27795	−0.006	65	35	30	38	25	70
48	15.27887	15.28163	0.018	70	30	35	60	25	65
64	20.37183	20.36364	−0.040	30	60	55	idler		70
64	20.37183	20.37037	−0.007	45	25	30	50	20	55
64	20.37183	20.37037	−0.007	60	50	45	55	30	75
64	20.37183	20.37551	0.018	70	40	35	60	25	65

Table 11.6. MOD worms: screw cutting changewheel combinations for a lathe with an 8TPI imperial leadscrew

	Worm			Mandrel	1st stud		2nd stud		Lead screw
MOD	Required TPI	Resulting TPI	% error	Driver	Driven	Driver	Driven	Driver	Driven
0.40	20.21268	20.20408	−0.043	70	45	35	55	30	75
0.40	20.21268	20.21053	−0.011	25	40	38	idler		60
0.40	20.21268	20.21380	0.006	70	40	35	50	21	65
0.40	20.21268	20.22222	0.047	20	65	45	idler		35
0.50	16.17014	16.16327	−0.043	70	30	35	55	20	60
0.50	16.17014	16.17000	−0.001	40	21	25	35	20	55
0.50	16.17014	16.17778	0.047	25	65	45	idler		35
0.50	16.17014	16.17778	0.047	30	70	75	idler		65
0.60	13.47512	13.46939	−0.043	35	75	70	idler		55
0.60	13.47512	13.47273	−0.018	55	30	40	38	20	65
0.60	13.47512	13.47500	−0.001	40	21	30	35	20	55
0.60	13.47512	13.48148	0.047	30	65	45	idler		35
0.70	11.55010	11.55000	−0.001	20	55	40	idler		21
0.70	11.55010	11.55000	−0.001	50	35	40	45	30	55
0.70	11.55010	11.55556	0.047	50	25	45	40	20	65
0.70	11.55010	11.55556	0.047	60	20	30	50	25	65
0.75	10.78009	10.77551	−0.043	35	60	70	idler		55

Worm				Mandrel	1st stud		2nd stud		Lead screw
MOD	Required TPI	Resulting TPI	% error	Driver	Driven	Driver	Driven	Driver	Driven
0.75	10.78009	10.77818	−0.018	55	30	50	38	20	65
0.75	10.78009	10.78000	−0.001	60	21	25	35	20	55
0.75	10.78009	10.78519	0.047	45	65	75	idler		70
0.80	10.10634	10.10204	−0.043	70	30	40	55	35	75
0.80	10.10634	10.10526	−0.011	25	60	38	idler		20
0.80	10.10634	10.10690	0.006	70	20	35	50	21	65
0.80	10.10634	10.10872	0.024	75	40	65	55	25	70
0.90	8.98341	8.97959	−0.043	70	55	35	idler		50
0.90	8.98341	8.98246	−0.011	45	20	38	40	25	60
0.90	8.98341	8.98333	−0.001	60	35	50	55	40	70
0.90	8.98341	8.98765	0.047	60	20	45	65	30	70
1.00	8.08507	8.08163	−0.043	70	55	35	idler		45
1.00	8.08507	8.08500	−0.001	50	21	40	35	20	55
1.00	8.08507	8.08552	0.006	70	20	35	40	21	65
1.00	8.08507	8.08889	0.047	45	20	40	35	25	65
1.25	6.46806	6.46465	−0.053	45	40	55	idler		50
1.25	6.46806	6.46800	−0.001	50	21	40	35	25	55
1.25	6.46806	6.46905	0.015	70	38	60	55	40	65
1.25	6.46806	6.47111	0.047	50	20	45	35	25	65
1.50	5.39005	5.38776	−0.043	70	55	35	idler		30
1.50	5.39005	5.38947	−0.011	75	20	38	40	25	60
1.50	5.39005	5.39000	−0.001	50	21	40	35	30	55
1.50	5.39005	5.39259	0.047	45	65	75	idler		35
1.75	4.62004	4.62000	−0.001	40	55	50	idler	21	
1.75	4.62004	4.62000	−0.001	75	35	50	45	40	55
1.75	4.62004	4.62222	0.047	60	20	45	30	25	65
1.75	4.62004	4.62222	0.047	30	65	75	idler		20
2.00	4.04254	4.04040	−0.053	45	50	55	idler		25
2.00	4.04254	4.04082	−0.043	70	20	40	45	35	55
2.00	4.04254	4.04211	−0.011	50	20	38	30	25	40
2.00	4.04254	4.04444	0.047	60	65	75	idler		35

CHAPTER 12
GEARS

GEARS: INVOLUTE FORM

This is shown in the drawing below (for abbreviations see below).

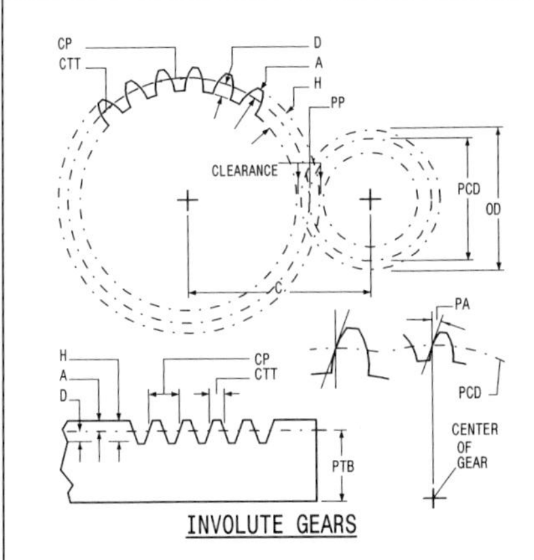

12-01: Gears: involute form

DEFINITIONS AND ABBREVIATIONS

Tooth size

There are three methods of defining gear tooth size:

Diametral pitch DP

The number of teeth per inch of pitch circle diameter.

Module MOD

The length, in millimeters, of the pitch circle diameter per tooth.

Circular pitch CP

The distance between adjacent teeth measured along the arc at the pitch circle diameter

Other definitions

Addendum A

The height of the tooth above the pitch circle diameter.

Center distance C

The distance between the axes of two gears in mesh.

Circular tooth thickness CTT

The width of a tooth measured along the arc at the pitch circle diameter.

Dedendum D

The depth of the tooth below the pitch circle diameter.

Outside diameter OD

The outside diameter of the gear.

Pitch circle diameter PCD

The diameter of the pitch circle.

Pitch point PP

The point at which the pitch circle diameters of two gears in mesh coincide. Effectively this is the diameters at which plain disks would create the same ratio if relying on friction alone.

Pitch to back PTB

The distance on a rack between the pitch circle diameter line and the rear face of the rack.

Pressure angle PA

The angle between the tooth profile at the pitch circle diameter and a radial line passing through the same point.

Whole depth H

The total depth of the space between adjacent teeth.

FORMULAS: MODULE (MOD) SYSTEM (METRIC)

Addendum:

$$A = \text{MOD}$$

Center distance:

$$C = \frac{PCD(g) + PCD(p)}{2}$$

where:

g = gear

p = pinion

Circular pitch:

$$CP = \pi \times \text{MOD}$$

Circular tooth thickness:

$$CTT = \frac{CP}{2}$$

Dedendum:

$$D = H - A$$

Module:

$$MOD = \frac{PCD}{N}$$

where:

N = number of teeth

Number of teeth:

$$N = \frac{PCD}{MOD}$$

Outside diameter:

$$OD = (N + 2) \times \text{MOD}$$

Pitch circle diameter:

$$PCD = N \times \text{MOD}$$

Whole depth (finer than 1.25MOD):

$$H = 2.4 \times \text{MOD}$$

Whole depth (1.25MOD and coarser)

$$H = 2.25 \times \text{MOD}$$

FORMULAS: DIAMETRAL PITCH (DP) SYSTEM (IMPERIAL)

Addendum:

$$A = \frac{1}{DP}$$

Center distance:

$$C = \frac{PCD(g) + PCD(p)}{2}$$

where:

g = gear

p = pinion

Circular pitch:

$$CP = \frac{\pi}{DP}$$

Circular tooth thickness:

$$CTT = \frac{CP}{2}$$

Dedendum:

$$D = H - A$$

Diametral pitch:

$$DP = \frac{N}{PCD}$$

Number of teeth:

$$N = DP \times PCD$$

Outside diameter:

$$OD = \frac{N + 2}{DP}$$

Pitch circle diameter:

$$PCD = \frac{N}{DP}$$

Whole depth (finer than 20DP):

$$H = \frac{2.4}{DP}$$

Whole depth (20DP and coarser):

$$H = \frac{2.25}{DP}$$

NOTES

1. In the formulas, $\pi = 3.1416$.

2. The pressure angle for commercially available gears is usually 20°. The drawing below shows (from left to right) a comparison of 20°, 30° and 10° pressure angle gear.

PRESSURE ANGLE COMPARISON

3. When two gears are meshed correctly their pitch circle diameters coincide at the pitch point (PP). A clearance then results between the top of the tooth on one gear and the bottom of the gap between two adjacent teeth on the other. The amount of clearance is the difference between the dedendum (D) and the addendum (A), i.e.:

Clearance = $D - A$.

4. When two gears mesh together, the larger one is called the gear and the smaller one the pinion. The gear that meshes with a rack is also called a pinion.

5. The shape of the space between adjacent teeth varies considerably according to the number of teeth on the gear. In gears with only a few teeth, the teeth are very rounded, whereas, in gears with a large number of teeth, they are almost straight-sided. Spaces between adjacent teeth that are cut by a milling cutter therefore vary considerably. In theory, a different cutter should be used according to the number of teeth. In practice, except perhaps for extremely critical applications, a compromise is reached in which 8 different cutters are used to cut from 12 teeth up to a rack. Table 12.1 indicates the range of sizes for each cutter. Note that metric cutters (MOD) are numbered in reverse.

6. Gears with small numbers of teeth (from 11 down to 6) require special consideration. Readers should refer to a specialist book on the subject before cutting gears of these sizes.

7. Commercially available gears with 16 or fewer teeth may have a modified tooth form known as an 'addendum modification' or 'corrected gears'. These mesh correctly with standard gears but at a modified center distance. The supplier should be able to provide the details.

8. The tooth shape of the rack is straight–sided, with an angle equal to the pressure angle.

9. The drawing on page 74, which is actual size and compares 24-tooth gears in the metric sizes 2 MOD, 1

MOD and 0.5 MOD. These are very similar to 12 DP, 24 DP and 48 DP, but the DP gears are slightly larger.

10. Table 12.2 compares the ranges of DP and MOD sizes. This table is not meant to imply any interchangeability or mixed usage, although for a few sizes, and in a non-demanding situation, this may be acceptable if gears are to hand.

11. A pair of gears, adequately lubricated, meshing smoothly, and at the correct center distance, should have a transmission efficiency in the order of 97%.

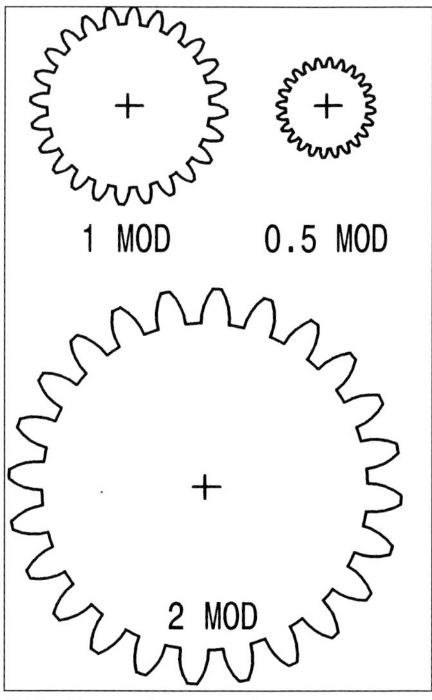

24-tooth gear size comparison

Table 12.1. Cutter numbers and ranges

DP no.	MOD no.	For cutting gears
1	8	135T to rack
2	7	55T to 134T
3	6	35T to 54T
4	5	26T to 34T
5	4	21T to 25T
6	3	17T to 20T
7	2	14T to 16T
8	1	12T to 13T

Table 12.2. Comparison of DP and MOD sizes

DP standard	MOD equivalent	MOD standard	DP equivalent
-	-	0.4	63.5
-	-	0.5	50.8
48	0.53	-	-
-	-	0.6	42.33
40	0.63	-	-
-	-	0.7	36.29
32	0.79	-	-
-	-	0.8	31.75
-	-	1.0	25.4
24	1.06	-	-
-	-	1.25	20.32
20	1.27	-	-
-	-	1.5	16.93
16	1.59	-	-
-	-	2.0	12.7
12	2.12	-	-
-	-	3.0	8.47
-	-	4.0	6.35

WORMS AND WORMWHEELS

TYPES

Mechanically the ideal combination is a wormwheel with a concave face that matches the diameter of the worm. As this is beyond what is possible for most home workshops, due to lack of ability, machinery or time, alternatives are often used.

Frequently the application for which they are required, e.g. a dividing head, does not have a very demanding duty, and these compromise arrangements are adequate.

METHODS

The three methods, in order of simplicity, but in reverse order of efficiency, are as follows.

1. A worm meshing with a standard spur gear and with the spindles at right angles (see drawing).

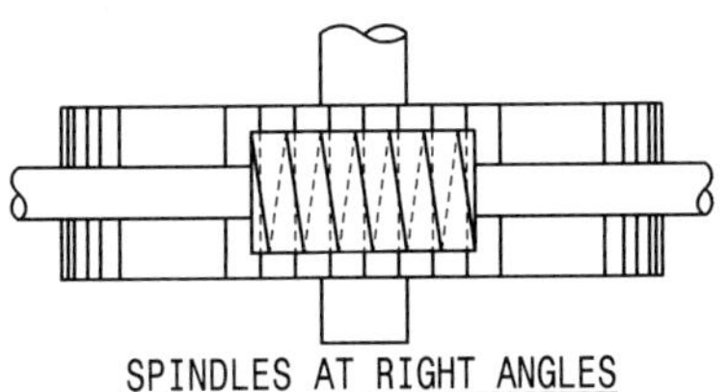

2. As for method 1, but with the two spindles angled to match the helix angle of the worm with the teeth on the spur gear (see drawing).

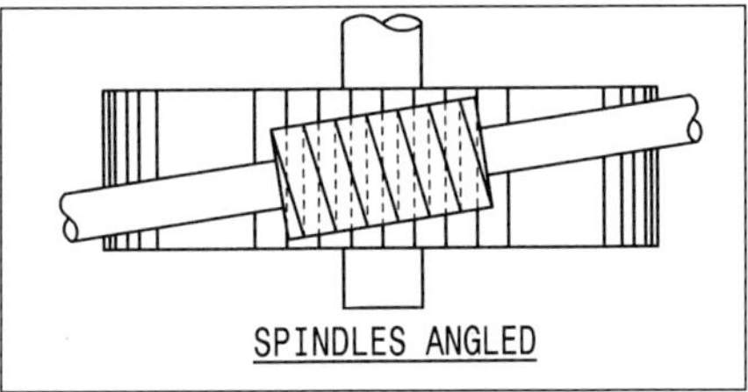

3. In this case the two spindles are at right angles and with the teeth on the wormwheel, generally as a standard spur gear, but angled to match the helix angle of the worm (see drawing).

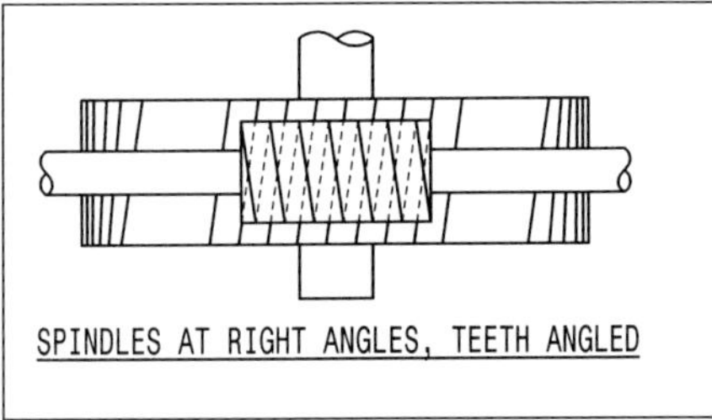

While method 1 is just about adequate for duties such as a dividing head, methods 2 and 3 are much better. These can be used for light-duty, low-speed, semi-continuous applications. Method 3 has the marked advantage that the spindles are at right angles, but it does require a special wormwheel.

With method 1, the largest possible worm diameter should be used because this limits its helix angle and therefore the mismatch between worm and wormwheel.

For methods 1 and 2, the worm may appear identical, but the angled worm has a marginal effect on its pitch that may or may not be taken into account (see 'Formulas' below).

FORMULAS

Method 1

In method 1, the pitch of the worm can be assumed to be equal to the circular pitch (CP) of the standard spur gear that it is used with, i.e.:

Worm pitch:

$$CP = \frac{\pi}{DP} \quad \text{or} \quad \pi \times MOD$$

This is the basis of the tables in Chapter 11.

Method 2

The effect of the angled worm is to increase the pitch of the worm, as shown in the drawing below. In this case, the formula is:

$$\text{Worm pitch} = \frac{CP}{\cos Ha}$$

where:

Ha = helix angle

As this adjusted pitch is dependent on the helix angle, which is itself dependent on the diameter of the worm, no actual values can be given and therefore no changewheel combinations are listed. However, where a reasonably large worm diameter is employed, the change is small. Using the coarsest pitch from the tables will probably be sufficient, although it may

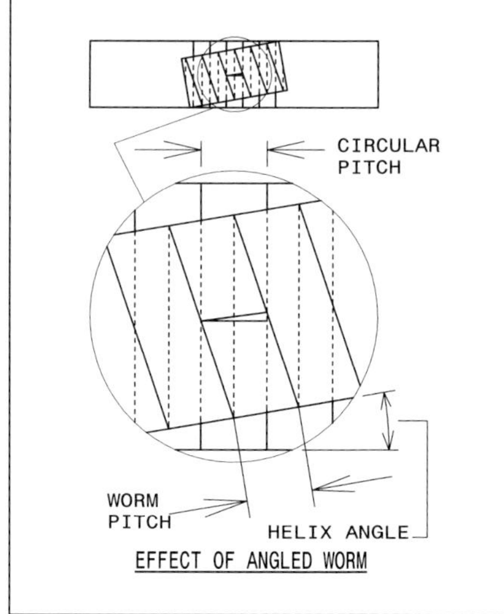

12-07

still be smaller than the theoretical pitch required.

Method 3

Determining the required dimensions is more complex than for methods 1 and 2, because, in theory at least, the angled teeth require the gear to be cut on a larger diameter blank.

The drawing, albeit exaggerated, shows how the thickness of the tooth (T) is narrower than normal if the diameter, and therefore the circular pitch (CP), are made the same as a conventional spur gear. The

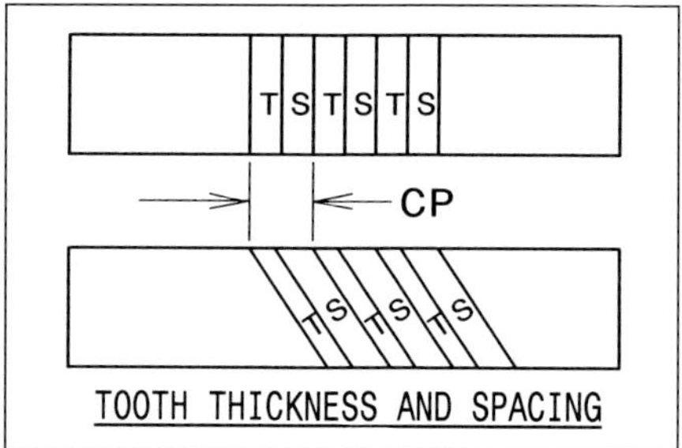

reason for this is that, although the space (S) between the teeth remains the same, because it is cut with the same cutter, due to the helix angle (Ha), there is less space left for the tooth.

Therefore, the circular pitch (CP) has to be increased (CPI) and, to achieve this, the pitch circle diameter (PCD) must be similarly increased (PCDI). The formulas are:

$$CPI = \frac{\pi}{DP \times \cos Ha} \quad or \quad \frac{\pi \times MOD}{\cos Ha}$$

and:

$$PCDI = \frac{N}{DP \times \cos HA} \quad or \quad \frac{N \times MOD}{\cos Ha}$$

The outside diameter (OD) is equal to the PCDI plus twice the addendum (MOD) i.e.:

$$OD = PCDI + \frac{2}{DP} \quad or \quad PCDI + (2 \times MOD)$$

where:

Ha = helix angle

$\pi = 3.1416$

See also pages 71–73.

WORMS

Unlike the diameter of a spur gear or wormwheel, which is fixed by the chosen tooth size and number of teeth, worm diameter can vary over a wide range. The only limit is that it must not be so small as to make the helix angle unacceptably large. However, the diameter can be increased to suit the mechanical requirements (spindle centers etc.) of the design. The center distance is calculated as follows:

$$C = \frac{PCD \, (worm) + PCD \, (wormwheel)}{2}$$

Once the PCD of the worm has been decided, its outside diameter (OD) can be found by adding twice the addendum to the PCD, i.e.:

$$OD = PCD + \frac{2}{DP} \quad or \quad PCD + (2 \times MOD)$$

PITCH ERRORS

Obtaining the exact worm pitch required by using the available changewheels is rare, although very close values are normally possible. Calculating these values manually can be difficult, but a dedicated computer program will produce them with ease. It should be noted that the error in pitch in no way affects the worm to wormwheel ratio, which is always equal to the number of teeth on the wormwheel.

HELIX ANGLE

The helix angle (Ha) is calculated from the worm's circumference at the pitch circle diameter ($\pi \times$ PCD) and its pitch (P). Note that it is the pitch circle diameter, not the outside diameter, that is used in this calculation:

$$\tan Ha = \frac{P}{\pi \times PCD}$$

CONVENTIONAL WORMS AND WORMWHEELS

As mentioned earlier, conventional worms and wormwheels are beyond the scope of this book and what can normally be produced in the small workshop. While the worms are like those above, the need to purchase or make a special hob to cut the wormwheels normally make this a non-starter. A gear hobbing machine is also required, or a lathe attachment to provide the same facilities.

Even so, making conventional worms and wormwheels is not beyond the scope of most home workshops, but anyone wishing to do so should seek further details (see page 187).

CHAPTER 13
BELT DRIVES

V-BELT AND WEDGE-BELT DRIVES

Two forms of belt drive are available, both frequently referred to as V-belt drives. However, one form is called a V-belt while the correct name for the other is a wedge belt. The V-belt is the most popular.

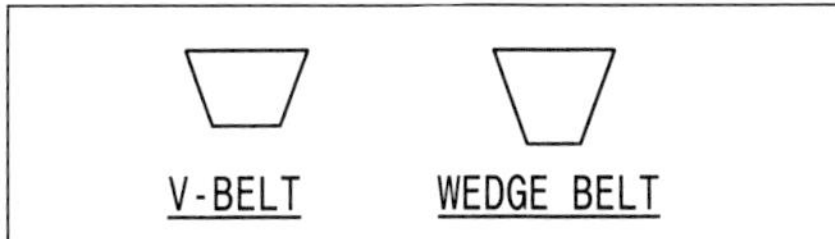

PITCH WIDTH AND DIAMETER

The ratio between two pulleys of differing diameters is worked out by using their effective diameters. The effective diameter (ED)is neither the outside diameter (OD) nor the diameter at the base of the grooves, but the diameter at some point down the groove. This is known as the pitch diameter (Dp). The width of the groove at this point is similarly called the pitch width (Wp). This point is also called the pitch width of the belt and is the point at which the length of the belt is measured.

BELT DIMENSIONS

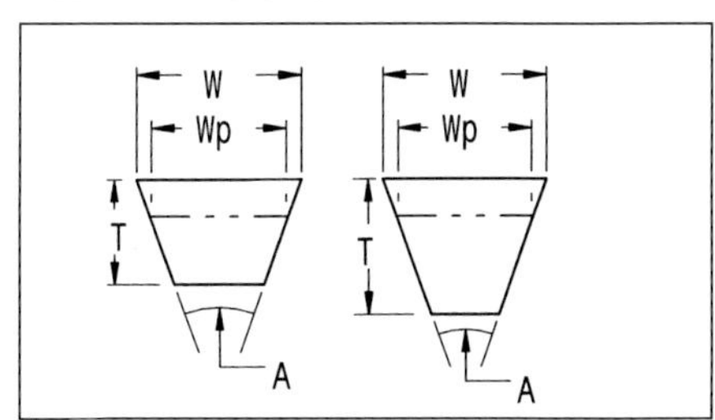

Table 13.1

Cross-section ref.	Wp	W	T	A
Wedge belts				
SPZ	8.5	10	8	40
SPA	11.0	13	10	40
SPB	14.0	17	14	40
SPC	19.0	22	18	40
V-belts				
Y	5.3	6	4	40
Z	8.5	10	6	40
A	11.0	13	8	40
B	14.0	17	11	40
C	19.0	22	14	40
D	27.0	32	19	40

PULLEY DIMENSIONS

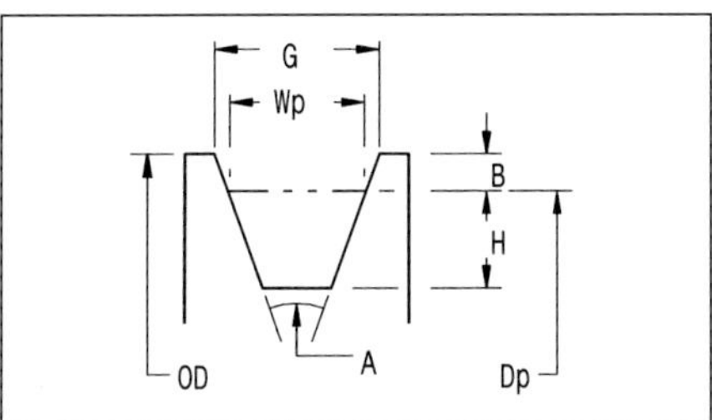

Table 13.2: All dimensions are in millimeters

Cross-section ref.	Wp	H (1/3)	B (3)	Dp (2)	A (3)	G (3)
Y	5.3	4.7	1.6	60−	32	6.2
-	-	-	-	60+	36	6.3
SPZ/Z	8.5	9.0/7.2	2.0	80−	34	9.7
-	-	-	-	80+	38	9.9

Cross-section ref.	Wp	H (1/3)	B (3)	Dp (2)	A (3)	G (3)
SPA/A	11.0	11.0/8.8	2.75	118–	34	12.7
-	-	-	-	118+	38	12.9
SPB/B	14.0	14.0/11.2	3.5	190–	34	16.1
-	-	-	-	190+	38	16.4
SPC/C	19.0	19.0/15.2	4.8	315–	34	21.9
-	-	-	-	315+	38	22.3
D	27.0	19.9	8.1	475–	36	32.3
-	-	-	-	475+	38	32.6

1. H: the smaller dimensions can be used for V-belts with references Z, A, B and C.

2. Dp (pitch diameter): the minus sign indicates values up to and including the value indicated, while the plus sign indicates values higher than the quoted value.

3. B, G and H are minimum values and A (angle of contact) is ± 0.5°. As a result, some pulley manufacturers may not work rigidly to the values for H and B. In these cases, it is not be possible to arrive at the dimension for the pitch diameter by measuring the pulley and applying either H or B.

PULLEY BELT LENGTH AND PULLEY CENTERS CALCULATIONS

To determine belt length (L):

$$L = 2C + \frac{(D - d)^2}{4C} + \frac{\pi(D + d)}{2}$$

To determine pulley centers (C):

$$C = A + \sqrt{A^2 - B}$$

where:

$$A = \frac{L}{4} - \frac{\pi(D + d)}{8}$$

and:

$$B = \frac{(D - d)^2}{8}$$

For all above formulas:

C = pulley center distance

D = pitch diameter of large pulley

d = pitch diameter of small pulley

L = pitch length of belt

$\pi = 3.1416$

To determine pitch length of an existing belt:

For V-belts:

$$Lp = Li + K$$

For wedge belts:

$$Lp = Lo - K$$

where:

Lp = pitch length

Li = inside length

Lo = outside length

K = constant for each belt size.

Value of K:

Table 13.3

V-belts		Wedge belts	
Ref.	K	Ref.	K
Y	15	-	-
Z	25	SPZ	13
A	33	SPA	17
B	43	SPB	22
C	62	SPC	26
D	82	-	-

All dimensions are in millimeters.

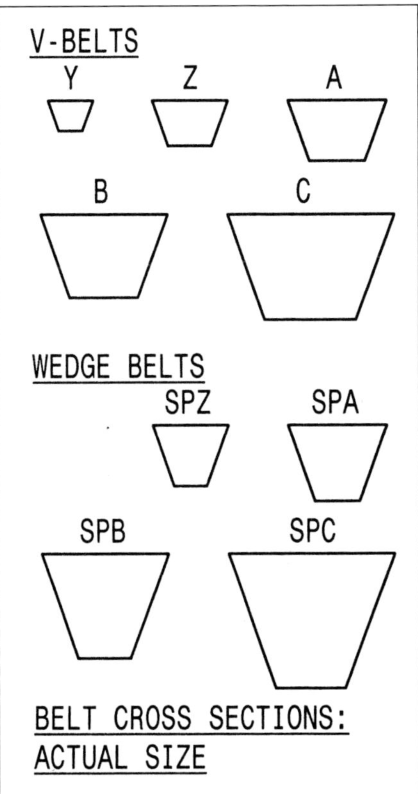

TOOTHED BELT DRIVES

Toothed belt drives are known by various names, mainly 'timing belt' or 'synchronous belt'. They provide precise ratios, as is possible with chain drives, but can run at higher speeds and are quieter and virtually maintenance-free. The two most common types (there are others) are the Classical range (predominantly imperial sizes) and the high torque drives (HTD, metric sizes), which have quite different tooth forms. Some suppliers use their own catalog references rather than the standard belt and pulley designations.

CLASSICAL RANGE

The Classical tooth form is rather rectangular compared to the rounded form of the HTD range. The range is made in six imperial sizes.

Coding

Pitch code

The six imperial sizes are coded according to pitch.

Pitch (in)	Code
0.08	MXL
0.2	XL
0.375	L
0.5	H
0.875	XH
1.25	XXH

Width code

The width code is the width in inches to two decimal places, but with the decimal point omitted, e.g., 0.75in is coded 075, and 3.00in is coded 300. The widths available depend on the pitch and are as follows:

Table 13.4.

Pitch code	Width code				
MXL	012	018	025	031	037
XL	025	031	037		
L	050	075	100		
H	075	100	150	200	300
XH	200	300	400		
XXH	200	300	400	500	

While this is basically an imperial belt, metric widths are available from some suppliers

Length code

Belts are made in a range of lengths but not in increments of one tooth (see suppliers' catalogs for lengths available). The length is equal to the number of teeth times the value for the pitch. Typically, a belt with 100 teeth and a pitch of 0.2in would have a pitch length of 20in. The code is the length rounded up to one decimal place but omitting the decimal point, i.e. 24.0in is coded as 240, and 12.375in is coded as 124.

Belt designation

The belt is designated by the length code, pitch code and width code, e.g. as 124XL031.

Belt: major dimensions

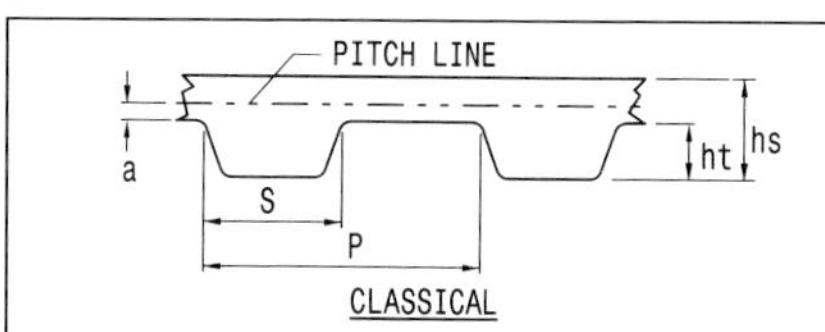

Table 13.5

Pitch code	P	S	ht	hs	a
MXL	0.080	–	0.02	0.045	0.010
XL	0.200	0.101	0.05	0.93	0.010
L	0.375	0.183	0.075	0.140	0.015
H	0.500	0.241	0.090	0.171	0.027
XH	0.875	0.495	0.250	0.440	–
XXH	1.250	0.750	0.375	0.620	–

Pulley designation

A pulley is designated by the number of teeth, pitch code and width code, e.g. as 72XL031.

HTD RANGE

The HTD tooth form is rounded and made to metric dimensions.

Coding

Pitch code

Pitch (mm)	Code
3	3M
5	5M
8	8M
14	14M

Width code

Width (mm)	Code
9	09
15	15
20	20
30	30
40	40
55	55

Length code

Comments for the Classical range regarding pitch length also apply to the HTD range, except that dimensions are in metric and there are no decimals, e.g., a belt with 70 teeth and a pitch of 5mm would have a pitch length of 350mm and a code of 350.

Belt designation

The belt is designated by the length code, pitch code and width code, e.g. as 350-5M-09.

Belt: major dimensions

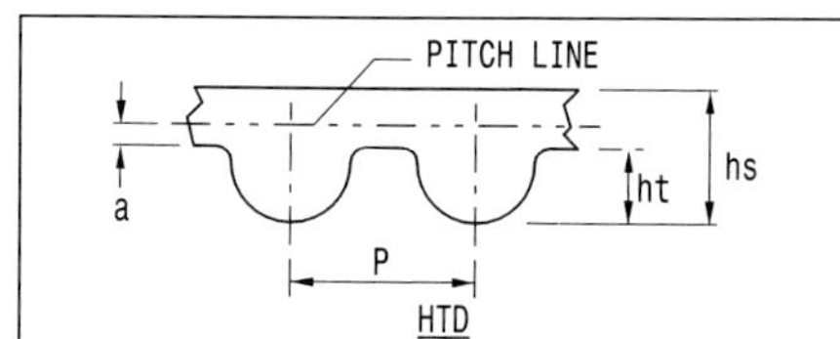

Table 13.6

Pitch code	P	ht	hs	a
3M	3	1.22	2.54	0.38
5M	5	2.06	3.81	0.57
8M	8	3.40	5.60	0.69
14M	14	6.10	10.00	1.38

Pulley designation

A pulley is designated by the number of teeth, pitch code and width code, e.g. as 48-5M-09. In a few cases, pulleys and belts can have the same code, so it is important to specify whether a pulley or a belt is required.

CENTER DISTANCE CALCULATIONS

The following formula is applicable to both Classical and HTD ranges. If pulleys are being purchased, the pitch circle diameters ('D' and 'd') should be obtainable from the suppliers' catalog. If existing pulleys are being used, the values for 'D' and 'd' can be determined by taking their outside diameters and adding twice the value of 'a' (taken from the tables above).

To determine center distance:

$$= \frac{K + \sqrt{K^2 - 32(D - d)^2}}{16}$$

where:

$K = 4L - 2\pi(D + d)$

D = pitch circle diameter of large pulley

d = pitch circle diameter of small pulley

L = belt length

The value obtained will be sufficiently accurate, provided that some adjustment is made for belt tensioning. For applications that will have no adjustment, contact the manufacturer for help in determining the center distance required.

RIBBED BELT DRIVES

These have the appearance of a flat belt with a number of small V-grooves along their length, these are often know as 'poly v' belts. The cross section does not have the same degree of wedge action as the larger V-belt so they have to work at a higher tension.

A major advantage of these belts is that they can work with smaller diameter pulleys than normal V belts. However, their power transmission rating is much reduced at these diameters. Also they will not slip from the pulley, as can happen with flat belting, and only a small amount of pulley movement is required to permit removal of the belts for speed changing.

SIZES

The belts are made in five profiles, referenced PH, PJ, PK, PL and PM. The pitch between each rib ranges from 1.6mm (PH) to9.4mm (PM).

BELT DESIGNATION

The belt is designated by the number of grooves, profile and effective length, e.g. as 10 PK 508.

BELT: MAJOR DIMENSIONS

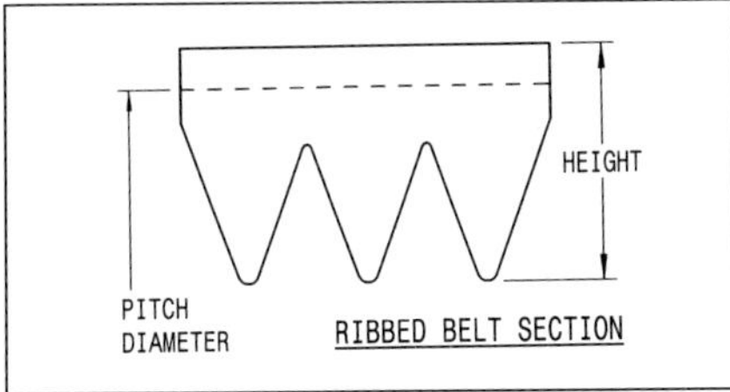

Belt size	Height (mm)
PH	3
PJ	4
PK	6
PL	10
PM	17

The width of the belt is nominally pitch times the number of grooves. This can be any number as the belt can be cut from a wider belt. However, the number will most likely be chosen to suit available pulleys, which are made in standard widths, i.e. 4 grooves, 8 grooves, etc.

PULLEY DESIGNATION

The pulley is designated by P, number of grooves, profile and the effective diameter, e.g. as P10PK150.

PULLEY: BASIC PROFILE

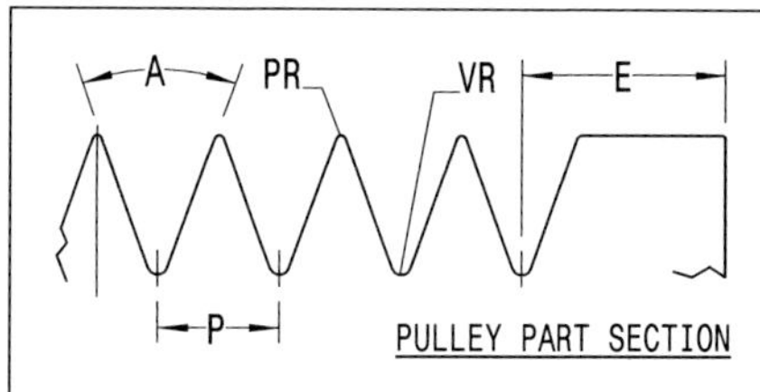

Angle A is 40° for all sizes.

Table 13.7

Profile	P	PR	VR	E
PH	1.60	0.15	0.3	1.3
PJ	2.34	0.20	0.4	1.8
PK	3.56	0.25	0.5	2.5
PL	4.70	0.40	0.4	3.3
PM	9.40	0.75	0.75	6.4

PULLEY: MODIFIED PROFILE

For ease of manufacture, deviation from the basic profile is permitted in the areas of PR and VR, as it is essential that the belt mates with the side faces of the V and not the radii. PR is not easy to make and can be replaced by a flat, thereby reducing the outside diameter. The resulting sharp edges must be removed to avoid damage to the belt. If this is done, the outside diameter (OD) will be reduced but, for calculations, the effective diameter (ED) must still be used.

The radius in the base of the groove (VR MOD)can also be reduced. In this case the form is optional, provided that it lies completely below the theoretical outline of VR.

Dimension E is a minimum and can be increased if it will benefit the design. Typically, a small-diameter pulley may be extended to provide space for a grub screw fixing.

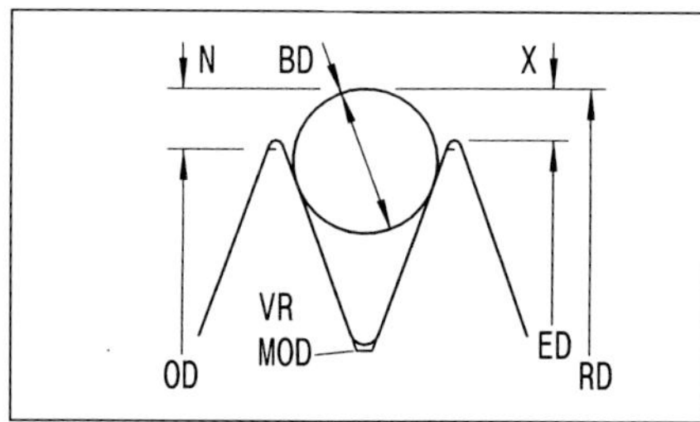

Table 13.8

Profile	BD	2X	2N	Min. diameter
PH	1.0	0.11	0.69	13
PJ	1.5	0.23	0.81	20
PK	2.5	0.99	1.68	45
PL	3.5	2.36	3.50	75
PM	7.0	4.53	5.92	180

CONSISTENT GROOVE DEPTH

In order to meet power ratings, the groove depths must be consistent with close limits. The previous drawing shows a ball or rod placed in one groove. One of these will be placed in the same groove on either side of the pulley and the dimension RD across them measured. The difference between the grooves is the important factor; the actual dimension is less critical as it has only a marginal effect on the drive ratio.

Dimension RD can also be used to arrive at the effective diameter (ED) for a pulley that has had its outside diameter (OD) reduced by making a flat.

To determine effective diameter (ED):

$$ED = RD - 2X$$

SPEED RATIO

The speed ratio is dependent on many factors, such as belt tension, power transmitted, etc. In practice, however, the following approach will be adequate for all normal applications. The effective pitch diameter of each pulley can be determined by adding the pitch adjustment (PA) to each pulley's effective diameter, the ratio then being the effective pitch diameter 1 divided by the effective pitch diameter 2.

Profile	PA (mm)
PH	1.6
PJ	2.4
PK	4.0
PL	6.0
PM	8.0

PULLEY BELT LENGTH AND PULLEY CENTERS CALCULATIONS

To determine belt length (L):

$$L = 2C + \frac{(D - d)^2}{4C} + \frac{\pi(D + d)}{2}$$

where:

C = pulley center
D = pitch diameter of large pulley
d = pitch diameter of small pulley
L = length of belt
$\pi = 3.1416$ for all formula in this section

To determine pulley centers (C):

$$C = A + \sqrt{A^2 - B}$$

where:

$$A = \frac{L}{4} - \frac{\pi(D + d)}{8}$$

$$B = \frac{(D - d)^2}{8}$$

D = pitch diameter of large pulley
d = pitch diameter of small pulley

The initial belt length is calculated using the desired pulley diameters and a center distance provisionally chosen to suit the design. The resulting belt length will almost certainly not conform to a standard, and a shorter or longer belt will then have to be chosen from the supplier's catalog. From this, the final pulley centers can then be calculated.

POWER RATING

As a guide to the main conditions affecting the power rating, the following are worth noting. The power rating is dependent on the diameter of the smallest pulley and increases considerably as the diameter increases, as shown in the following examples:

Diameter (mm)	kW/rib
20	0.01
40	0.15
60	0.29
80	0.41
100	0.53

As calculations are based on the smallest pulley, the arc of contact will be less than 180° in most cases, unless a jockey pulley is included in the design. This can reduce the power rating appreciably due to the reduction of contact with the pulley. Typically, at a 120° contact arc, the value is reduced by a factor of 0.8 when compared to a 180° contact arc.

The power rating increases with pulley speed, being only very marginal at smaller diameters, but almost proportional to speed at larger diameters.

The rating will be reduced for all but the lightest of duties, the following having a reducing effect:

- The use of high-starting torque motors, such as AC DOL or DC Series motors
- Frequent starting and stopping
- Intermittent duties, as seen with machine tools, particularly power presses
- Continuous running over many hours.
- The cumulative effect of these factors can result in a reduction in power rating by as much as 0.5.

As these comments show, the subject is very complex and users who are not completely confident that a design is well within limits are advised to consult the supplier of the drive belts for further information.

CHAPTER 14
DIVIDING

DIVIDING HEADS

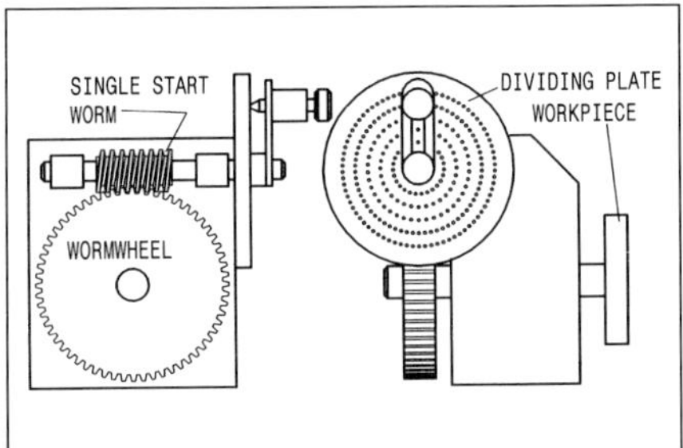

The drawing shows the basic construction of a dividing head but within this basis there are variations in the gear ratio and the number of holes on the dividing plates.

DIVISIONS

Tables 14.1 to 14.3 show the divisions possible using the commonly available ratios of 40:1, 60:1 and 90:1. Although 40:1 appears to be the standard for small commercial dividing heads, rotary tables, which can also have dividing plates fitted, may be either 60:1 or 90:1, hence their inclusion in the tables.

The manual supplied with a commercial head will quote just one set-up for each value, on the basis that the division plates supplied will remain with the head. However, this may not be the case and one plate may sometimes go missing. For many values this will not be a problem as the required division can be achieved by using another plate. For this reason, these tables include every set-up that provides an achievable division, e.g. Table 14.1 (for a 40:1 ratio) lists six set-ups for 24 divisions.

These tables can only cover a limited range of division plate numbers and are based on the plates commonly supplied with the smaller head. These numbers are 15, 16, 17, 18, 19, 20, 21, 23, 27, 29, 31, 33, 37, 39, 41, 43, 47 and 49.

For reasons of space, some values have been omitted. Values below 20 are not included because they can easily be arrived at by observation and some simple mathematics. Values above 200, with the exception of 360, have been omitted because of their limited application.

In Table 14.1 (40:1 head), no values are given for either 20 or 40 divisions. This is because, for 40 divisions, one turn of the input will rotate the output one division, and this is obviously achievable with any plate value. Similarly, 20 can be achieved with two full turns per division. For the same reason 20, 30 and 60 are omitted from Table 14.2 (60:1) and 30, 45 and 90 from Table 14.3 (90:1).

In each table, the column headings refer to the following:

D	Division achieved
P	Number of holes on the division plate to be used
T	Number of full turns at the input
H	Number of additional holes to be traversed after the number of full turns, if any.

Table 14.1. Divisions possible using a 40:1 ratio dividing head

D	P	T	H
21	21	1	19
22	33	1	27
23	23	1	17
24	15	1	10
24	18	1	12
24	21	1	14
24	27	1	18
24	33	1	22
24	39	1	26
25	15	1	9
25	20	1	12
26	39	1	21
27	27	1	13
28	21	1	9
28	49	1	21
29	29	1	11
30	15	1	5
30	18	1	6
30	21	1	7
30	27	1	9
30	33	1	11
30	39	1	13
31	31	1	9
32	16	1	4
32	20	1	5
33	33	1	7
34	17	1	3
35	21	1	3
35	49	1	7
36	18	1	2
36	27	1	3
37	37	1	3
38	19	1	1
39	39	1	1
41	41	0	40
42	21	0	20
43	43	0	40
44	33	0	30
45	18	0	16
45	27	0	24
46	23	0	20
47	47	0	40
48	18	0	15
49	49	0	40
50	15	0	12
50	20	0	16
52	39	0	30
54	27	0	20

D	P	T	H
55	33	0	24
56	21	0	15
56	49	0	35
58	29	0	20
60	15	0	10
60	18	0	12
60	21	0	14
60	27	0	18
60	33	0	22
60	39	0	26
62	31	0	20
64	16	0	10
65	39	0	24
66	33	0	20
68	17	0	10
70	21	0	12
70	49	0	28
72	18	0	10
72	27	0	15
74	37	0	20
75	15	0	8
76	19	0	10
78	39	0	20
80	16	0	8
80	18	0	9
80	20	0	10
82	41	0	20
84	21	0	10
85	17	0	8
86	43	0	20
88	33	0	15
90	18	0	8
90	27	0	12
92	23	0	10
94	47	0	20
95	19	0	8
98	49	0	20
100	15	0	6
100	20	0	8
104	39	0	15
105	21	0	8
108	27	0	10
110	33	0	12
115	23	0	8
116	29	0	10
120	15	0	5
120	18	0	6

D	P	T	H
120	21	0	7
120	27	0	9
120	33	0	11
120	39	0	13
124	31	0	10
128	16	0	5
130	39	0	12
132	33	0	10
135	27	0	8
136	17	0	5
140	21	0	6
140	49	0	14
144	18	0	5
145	29	0	8
148	37	0	10
150	15	0	4
152	19	0	5
155	31	0	8
156	39	0	10
160	16	0	4
160	20	0	5
164	41	0	10
165	33	0	8
168	21	0	5
170	17	0	4
172	43	0	10
180	18	0	4
180	27	0	6
184	23	0	5
185	37	0	8
188	47	0	10
190	19	0	4
195	39	0	8
196	49	0	10
200	15	0	3
200	20	0	4
360	18	0	2
360	27	0	3

Table 14.2. Divisions possible using a 60:1 ratio dividing head

D	P	T	H
21	21	2	18
21	49	2	42
22	33	2	24
23	23	2	14
24	16	2	8
24	18	2	9
24	20	2	10
25	15	2	6
25	20	2	8
26	39	2	12
27	18	2	4
27	27	2	6
28	21	2	3
28	49	2	7
29	29	2	2
31	31	1	29
32	16	1	14
33	33	1	27
34	17	1	13
35	21	1	15
35	49	1	35
36	15	1	10
36	18	1	12
36	21	1	14
36	27	1	18
36	33	1	22
36	39	1	26
37	37	1	23
38	19	1	11
39	39	1	21
40	16	1	8
40	18	1	9
40	20	1	10
41	41	1	19
42	21	1	9
42	49	1	21
43	43	1	17
44	33	1	12
45	15	1	5
45	18	1	6
45	21	1	7
45	27	1	9
45	33	1	11
45	39	1	13
46	23	1	7
47	47	1	13

D	P	T	H
48	16	1	4
48	20	1	5
49	49	1	11
50	15	1	3
50	20	1	4
51	17	1	3
52	39	1	6
54	18	1	2
54	27	1	3
55	33	1	3
57	19	1	1
58	29	1	1
62	31	0	30
63	21	0	20
64	16	0	15
65	39	0	36
66	33	0	30
68	17	0	15
69	23	0	20
70	21	0	18
70	49	0	42
72	18	0	15
74	37	0	30
75	15	0	12
75	20	0	16
76	19	0	15
78	39	0	30
80	16	0	12
80	20	0	15
81	27	0	20
82	41	0	30
84	21	0	15
84	49	0	35
85	17	0	12
86	43	0	30
87	29	0	20
90	15	0	10
90	18	0	12
90	21	0	14
90	27	0	18
90	33	0	22
90	39	0	26
92	23	0	15
93	31	0	20
94	47	0	30
95	19	0	12
96	16	0	10
98	49	0	30

D	P	T	H
99	33	0	20
100	15	0	9
100	20	0	12
102	17	0	10
105	21	0	12
105	49	0	28
108	18	0	10
108	27	0	15
110	33	0	18
111	37	0	20
114	19	0	10
115	23	0	12
116	29	0	15
117	39	0	20
120	16	0	8
120	18	0	9
120	20	0	10
123	41	0	20
124	31	0	15
126	21	0	10
129	43	0	20
130	39	0	18
132	33	0	15
135	18	0	8
135	27	0	12
138	23	0	10
140	21	0	9
140	49	0	21
141	47	0	20
145	29	0	12
147	49	0	20
148	37	0	15
150	15	0	6
150	20	0	8
155	31	0	12
156	39	0	15
160	16	0	6
162	27	0	10
164	41	0	15
165	33	0	12
170	17	0	6
172	43	0	15
174	29	0	10
180	15	0	5
180	18	0	6
180	21	0	7
180	27	0	9
180	33	0	11

D	P	T	H
180	39	0	13
185	37	0	12
186	31	0	10
188	47	0	15
190	19	0	6
192	16	0	5
195	39	0	12
196	49	0	15
198	33	0	10
200	20	0	6
360	18	0	3

Table 14.3. Divisions possible using a 90:1 ratio dividing head

D	P	T	H
20	16	4	8
20	18	4	9
20	20	4	10
21	21	4	6
21	49	4	14
22	33	4	3
23	23	3	21
24	16	3	12
24	20	3	15
25	15	3	9
25	20	3	12
26	39	3	18
27	15	3	5
27	18	3	6
27	21	3	7
27	27	3	9
27	33	3	11
27	39	3	13
29	29	3	3
31	31	2	28
32	16	2	13
33	33	2	24
34	17	2	11
35	21	2	12
35	49	2	28
36	16	2	8
36	18	2	9
36	20	2	10
37	37	2	16
38	19	2	7
39	39	2	12
40	16	2	4
40	20	2	5
41	41	2	8

D	P	T	H
42	21	2	3
42	49	2	7
43	43	2	4
46	23	1	22
47	47	1	43
48	16	1	14
49	49	1	41
50	15	1	12
50	20	1	16
51	17	1	13
54	15	1	10
54	18	1	12
54	21	1	14
54	27	1	18
54	33	1	22
54	39	1	26
55	33	1	21
57	19	1	11
58	29	1	16
60	16	1	8
60	18	1	9
60	20	1	10
62	31	1	14
63	21	1	9
63	49	1	21
65	39	1	15
66	33	1	12
69	23	1	7
70	21	1	6
70	49	1	14
72	16	1	4
72	20	1	5
74	37	1	8
75	15	1	3
75	20	1	4
78	39	1	6
80	16	1	2
81	18	1	2
81	27	1	3
82	41	1	4
85	17	1	1
86	43	1	2
87	29	1	1
93	31	0	30
94	47	0	45
95	19	0	18
96	16	0	15
98	49	0	45

D	P	T	H
99	33	0	30
100	20	0	18
102	17	0	15
105	21	0	18
105	49	0	42
108	18	0	15
110	33	0	27
111	37	0	30
114	19	0	15
115	23	0	18
117	39	0	30
120	16	0	12
120	20	0	15
123	41	0	30
126	21	0	15
126	49	0	35
129	43	0	30
130	39	0	27
135	15	0	10
135	18	0	12
135	21	0	14
135	27	0	18
135	33	0	22
135	39	0	26
138	23	0	15
141	47	0	30
144	16	0	10
145	29	0	18
147	49	0	30
150	15	0	9
150	20	0	12
153	17	0	10
155	31	0	18
160	16	0	9
162	18	0	10
162	27	0	15
165	33	0	18
170	17	0	9
171	19	0	10
174	29	0	15
180	16	0	8
180	18	0	9
180	20	0	10
185	37	0	18
186	31	0	15
189	21	0	10

D	P	T	H
190	19	0	9
195	39	0	18
198	33	0	15
200	20	0	9
360	16	0	4
360	20	0	5

Sometimes, dividing heads are produced within the workshop using the lathe's changewheels as a wormwheel. This opens up possibilities for other divisions. Table 14.4 lists all the additional divisions that can be achieved, although most are of limited use. However, one of them, 125, would be required if calibrating a dial for an 8TPI leadscrew.

The additional column 'R' refers to the ratio, and therefore the size, of the gear being used, e.g. a 55-tooth gear gives a 55:1 ratio.

Table 14.4. Additional divisions

D	P	T	H	R
77	21	0	15	55
77	33	0	30	70
77	33	0	15	35
77	49	0	35	55
91	21	0	15	65
91	39	0	30	70
91	39	0	15	35
91	49	0	35	65
112	16	0	5	35
112	16	0	10	70
119	17	0	10	70
119	17	0	5	35
121	33	0	15	55
125	15	0	3	25
125	15	0	6	50
125	15	0	9	75
125	20	0	4	25
125	20	0	8	50
125	20	0	12	75
133	19	0	5	35
133	19	0	10	70

D	P	T	H	R
143	33	0	15	65
143	39	0	15	55
153	17	0	5	45
154	33	0	15	70
161	23	0	5	35
161	23	0	10	70
169	39	0	15	65
171	19	0	5	45
175	15	0	3	35
175	15	0	6	70
175	20	0	4	35
175	20	0	8	70
175	21	0	6	50
175	21	0	3	25
175	21	0	9	75
175	49	0	7	25
175	49	0	14	50
175	49	0	21	75
176	16	0	5	55
182	39	0	15	70
187	17	0	5	55
189	21	0	5	45
189	27	0	5	35
189	27	0	10	70

CHAPTER 15

PRESSWORK

BENDING ALLOWANCE

The material in the outer part of any bend will be under tension , while the material in the inner part will be under compression. The position of the neutral axis (path of no change in length) will depend on the relative strengths of the material under compression and tension.

The bending allowances given in the tables are based on the neutral axis being at a position of 0.4 times the material thickness (t) and nearest the inside.

For a 90° bend, the bending allowance (C) therefore equals:

$$C = \frac{\pi \times 2 \times (0.4t + R)}{4}$$

$$= 1.57(0.4t + R)$$

where:

R = radius of inside bend

t = thickness of material

$\pi = 3.1416$.

The value of 0.4 should prove adequate for most materials. However, in critical applications, a trial should be carried out on a test sample.

Length of material equals:

A + B + C.

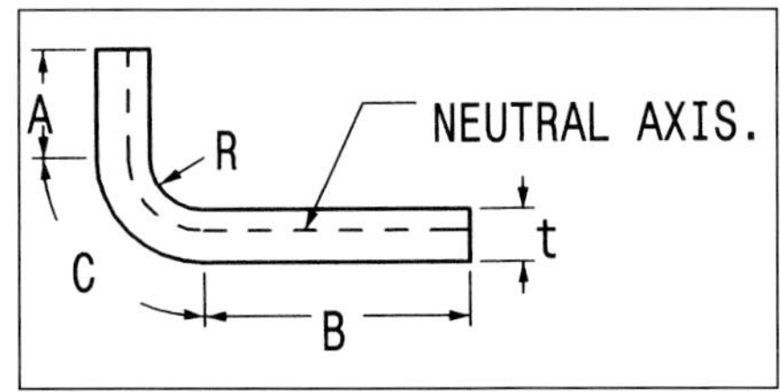

Bend allowance dimensions

For bends other than 90°, the bending allowance equals:

$$\frac{C \times A}{90}$$

where:

C = bending allowance (as per tables)

A = angle of bend

BENDING ALLOWANCE TABLES

The tabular matter overleaf gives bending allowance values for metric, imperial and SWG materials for a range of thicknesses and internal bend radii.

Table 15.1 Bending allowance: metric sizes

Thickness (mm)	Radius (mm)									
	1.0	2.0	3.0	4.0	5.0	6.0	7.0	8.0	9.0	10.0
0.5	1.9	3.5	5.0	6.6	8.2	9.7	11.3	12.9	14.5	16.0
0.6	1.9	3.5	5.1	6.7	8.2	9.8	11.4	12.9	14.5	16.1
0.7	2.0	3.6	5.2	6.7	8.3	9.9	11.4	13.0	14.6	16.1
0.8	2.1	3.6	5.2	6.8	8.4	9.9	11.5	13.1	14.6	16.2
0.9	2.1	3.7	5.3	6.8	8.4	10.0	11.6	13.1	14.7	16.3
1.0	2.2	3.8	5.3	6.9	8.5	10.1	11.6	13.2	14.8	16.3
1.1	-	3.8	5.4	7.0	8.5	10.1	11.7	13.3	14.8	16.4
1.2	-	3.9	5.5	7.0	8.6	10.2	11.7	13.3	14.9	16.5
1.3	-	4.0	5.5	7.1	8.7	10.2	11.8	13.4	15.0	16.5
1.4	-	4.0	5.6	7.2	8.7	10.3	11.9	13.4	15.0	16.6
1.5	-	4.1	5.7	7.2	8.8	10.4	11.9	13.5	15.1	16.7
1.6	-	4.1	5.7	7.3	8.9	10.4	12.0	13.6	15.1	16.7
1.7	-	4.2	5.8	7.4	8.9	10.5	12.1	13.6	15.2	16.8
1.8	-	4.3	5.8	7.4	9.0	10.6	12.1	13.7	15.3	16.8
1.9	-	4.3	5.9	7.5	9.0	10.6	12.2	13.8	15.3	16.9
2.0	-	4.4	6.0	7.5	9.1	10.7	12.3	13.8	15.4	17.0
2.1	-	-	6.0	7.6	9.2	10.7	12.3	13.9	15.5	17.0
2.2	-	-	6.1	7.7	9.2	10.8	12.4	13.9	15.5	17.1
2.3	-	-	6.2	7.7	9.3	10.9	12.4	14.0	15.6	17.2
2.4	-	-	6.2	7.8	9.4	10.9	12.5	14.1	15.6	17.2
2.5	-	-	6.3	7.9	9.4	11.0	12.6	14.1	15.7	17.3
2.6	-	-	6.3	7.9	9.5	11.1	12.6	14.2	15.8	17.3
2.7	-	-	6.4	8.0	9.6	11.1	12.7	14.3	15.8	17.4
2.8	-	-	6.5	8.0	9.6	11.2	12.8	14.3	15.9	17.5
2.9	-	-	6.5	8.1	9.7	11.2	12.8	14.4	16.0	17.5
3.0	-	-	6.6	8.2	9.7	11.3	12.9	14.5	16.0	17.6
3.5	-	-	-	8.5	10.1	11.6	13.2	14.8	16.3	17.9
4.0	-	-	-	8.8	10.4	11.9	13.5	15.1	16.7	18.2
4.5	-	-	-	-	10.7	12.3	13.8	15.4	17.0	18.5
5.0	-	-	-	-	11.0	12.6	14.1	15.7	17.3	18.8
5.5	-	-	-	-	-	12.9	14.5	16.0	17.6	19.2
6.0	-	-	-	-	-	13.2	14.8	16.3	17.9	19.5
6.5	-	-	-	-	-	-	15.1	16.7	18.2	19.8

Table 15.2 Bending allowance: imperial sizes

Thickness (in)		Radius (in)									
Decimal	Fraction	0.063	0.094	0.125	0.157	0.187	0.219	0.250	0.312	0.375	0.500
		1/16	3/32	1/8	5/32	3/16	7/32	1/4	5/16	3/8	1/2
0.032	1/32	0.12	0.17	0.22	0.27	0.31	0.36	0.41	0.51	0.61	0.81
0.063	1/16	0.14	0.19	0.24	0.29	0.33	0.38	0.43	0.53	0.63	0.82
0.094	3/32	-	0.21	0.26	0.31	0.35	0.40	0.45	0.55	0.65	0.84
0.125	1/8	-	-	0.27	0.33	0.37	0.42	0.47	0.57	0.67	0.86
0.157	5/32	-	-	-	0.35	0.39	0.44	0.49	0.59	0.69	0.88
0.187	3/16	-	-	-	-	0.41	0.46	0.51	0.61	0.71	0.90
0.219	7/32	-	-	-	-	-	0.48	0.53	0.63	0.73	0.92
0.250	1/4	-	-	-	-	-	-	0.55	0.65	0.75	0.94
0.313	5/16	-	-	-	-	-	-	-	0.69	0.79	0.98
0.375	3/8	-	-	-	-	-	-	-	0.73	0.82	1.02

Table 15.3 Bending allowance: SWG

Thickness		Radius (in)									
SWG	in	0.032	0.063	0.094	0.125	0.157	0.187	0.219	0.250	0.312	0.375
		1/32	1/16	3/32	1/8	5/32	3/16	7/32	1/4	5/16	3/8
22	0.028	0.07	0.12	0.16	0.21	0.26	0.31	0.36	0.41	0.51	0.61
21	0.032	0.07	0.12	0.17	0.22	0.27	0.31	0.36	0.41	0.51	0.61
20	0.036	-	0.12	0.17	0.22	0.27	0.32	0.37	0.42	0.51	0.61
19	0.040	-	0.12	0.17	0.22	0.27	0.32	0.37	0.42	0.52	0.61
18	0.048	-	0.13	0.18	0.23	0.28	0.32	0.37	0.42	0.52	0.62
17	0.056	-	0.13	0.18	0.23	0.28	0.33	0.38	0.43	0.53	0.62
16	0.064	-	0.14	0.19	0.24	0.29	0.33	0.38	0.43	0.53	0.63
15	0.072	-	-	0.19	0.24	0.29	0.34	0.39	0.44	0.54	0.63
14	0.080	-	-	0.20	0.25	0.30	0.34	0.39	0.44	0.54	0.64
13	0.092	-	-	0.20	0.25	0.30	0.35	0.40	0.45	0.55	0.65
12	0.104	-	-	-	0.26	0.31	0.36	0.41	0.46	0.56	0.65
11	0.116	-	-	-	0.27	0.32	0.37	0.42	0.47	0.56	0.66
10	0.128	-	-	-	0.28	0.33	0.37	0.42	0.47	0.57	0.67
9	0.144	-	-	-	-	0.34	0.38	0.43	0.48	0.58	0.68
8	0.160	-	-	-	-	0.35	0.39	0.44	0.49	0.59	0.69
7	0.176	-	-	-	-	-	0.40	0.45	0.50	0.60	0.70
6	0.192	-	-	-	-	-	0.41	0.46	0.51	0.61	0.71
5	0.212	-	-	-	-	-	-	0.48	0.53	0.62	0.72
4	0.232	-	-	-	-	-	-	-	0.54	0.64	0.73
3	0.252	-	-	-	-	-	-	-	0.55	0.65	0.75

HOLE PUNCHING

PRESSURE TO PUNCH

The pressure required to punch a hole in a flat plate, metal, plastic, card, etc. is dependent on the following:

1. Distance around the hole's periphery

2. Thickness of the material

3. Type of material

4. Clearance between punch and die

5. Presence or absence of lubrication

6. The effects of 4 and 5 are slight, and less than variations due to differing grades of materials. They can therefore be ignored.

TONNAGE

tonnage
= hole perimeter × material thickness × MC
where:
MC = material constant
Perimeter and thickness are given in millimeters.

The following material constants (MC) are accurate enough for normal applications.

Aluminum (hard)	0.02
Brass (hard)	0.03
Copper	0.02
Mild steel	0.04
Stainless steel	0.06

Materials purchased soft, or annealed by the user, will reduce the MC by about 25%.

EXAMPLE

To punch a hole of 15mm diameter through 0.8mm-thick hard aluminum, the pressure required is:

15 × 3.14 × 0.8 × 0.02 =0.75 tons

This is just within the capacity of the larger lever presses, but is well within the capacity of even the smallest fly press, which provides a pressure of about 3 tons.

CLEARANCE

A clearance between the punch and die is required, the amount depending on the thickness of the material being punched. With the clearance expressed as the difference in the diameters of the punch and die, it should be about 0.1 times the material thickness. The amount is not critical, and if a punch and die giving a larger clearance is all that is available these can be tried. However, thin materials will require punch-and-die combinations with minimal clearance.

PUNCHING AND BLANKING

The hole produced is equal to the punch diameter, while the disk produced is equal to the die diameter. In critical applications, the punch should be to size if the hole is important, and the die should be to size if the blank is important.

ANGULAR CLEARANCE

In both the above cases the die should be tapered slightly, e.g. 3°, after a short parallel portion about 4mm deep. This is to ease the blanks through the die. Ideally, even the parallel portions should be tapered very slightly, although in most cases this can be ignored.

CHAPTER 16

WELDING SYMBOLS

The diagrams used in industry invariably indicate welding requirements by the use of symbols that conform to set standards. In some cases, these may find their way into the small workshop and, for this reason, some explanation of the system is appropriate.

A full explanation would be overly long, but the details given are sufficient to enable the reader to understand how they relate to welding. The examples given probably cover the most common symbols and will give an adequate understanding of the system for most situations. However, if a project is particularly demanding, the full standard, which is far more detailed, should be consulted. In addition, national standards can vary from one country to another.

The symbols used typically define:

- the position of the weld
- the size of the weld
- whether the weld is continuous or has gaps
- whether the weld is to be dressed flat or left raised
- what preparation of the workpiece, e.g. chamfering, is to be carried out.

EXAMPLES

The symbols, and how they are used, are best understood by reference to the examples shown in the following series of drawings. These each include an illustration of the weld in question and a working drawing showing the symbols.

POSITION OF THE WELD

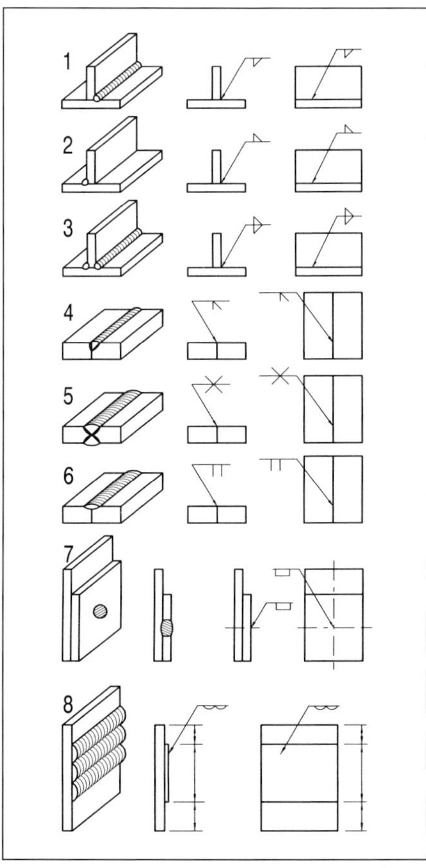

The first three examples apply to a basic fillet. Note the position of the symbols (the triangles) relative to the horizontal (symbol) lines on which they are drawn.

Example 1: In this case the triangle is below the horizontal line, indicating that the weld is on the side indicated by the arrow.

Example 2: In this case the triangle is above the horizontal line, indicating that the weld is on the opposite side to that indicated by the arrow (compare examples 1 and 2). See also 'Alternative system' below.

Example 3: When there is a weld on both sides, the triangle is repeated on both sides of the horizontal line.

PREPARATION

Where two pieces of metal are butted together the symbol may also include some indication of the required preparation of the parts prior to welding.

Example 4: This shows two pieces of metal that are welded together on one side only, and with one of the two pieces chamfered prior to welding. Note that the symbol consists of one line at 90° to the horizontal line and one angled line. This indicates that only one part is chamfered.

Example 5: If both parts are to be chamfered then both lines are angled. In this case, the symbol is repeated above and below the line, showing that preparation and welding is to be carried out on both sides of the parts.

Example 6: This shows two parts butted together without any chamfering. This is indicated by two lines at 90° to the horizontal line. As the symbols are below the horizontal line, the weld is only on the side indicated by the arrow.

Example 7: This shows two parts welded together using a plug weld made through a pre-prepared hole in one part only.

Example 8: Sometimes a weld is used to build up a surface, as in this example, which shows the dimensions that are required to detail the area and depth of the built-up surface.

FINISHING

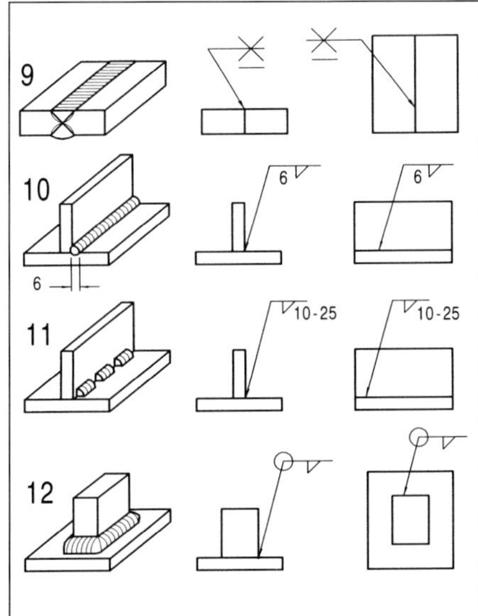

In some cases it is necessary to machine the weld and this is indicated by supplementary symbols. A straight line added to a symbol indicates that the weld should be finished flush.

Example 9: In this example, both sides are welded but only one side is machined flush (compare this with example 5). A letter is sometimes added to indicate the method of machining, e.g. 'G' for grinding. In some cases the straight line beneath the symbols will be replaced by a curved line. Depending on which way up this is, it will indicate either a concave or a convex finish.

DIMENSIONS

In many cases it is necessary to add dimensions for the cross section and length of the weld.

Example 10: The number to the left of the symbol indicates the weld cross section. A number to the right would indicate its length.

Example 11: Two numbers to the right of the symbol indicate short lengths of weld repeated along the length of the part, with the first number being the length of the welds and the second number the centers between them.

Example 12: In some cases a stud or some other projection has to be welded to a flat surface. This may require a fillet weld around its complete circumference. In this, or in any similar situation, the requirement to weld completely round an item is indicated by a circle drawn where the arrow line meets the horizontal line.

ALTERNATIVE SYSTEM

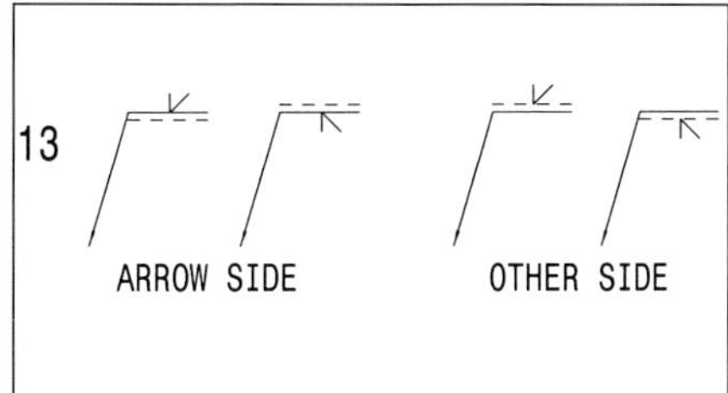

Example 13: This uses a solid line with a parallel broken line. In this case, if the symbol is drawn on the solid line, the weld is on the same side as the arrow, while if it is drawn on the broken line it is on the opposite side. The broken line can be either side of the solid line, and the two lines can be either horizontal or vertical.

CHAPTER 17
MATHEMATICAL FORMULAS

Note: For all formula π (pi) = 3.1416

CIRCLE

Circumference = $\pi \times D$

$$\text{Area} = \frac{\pi \times D^2}{4}$$

SEGMENT

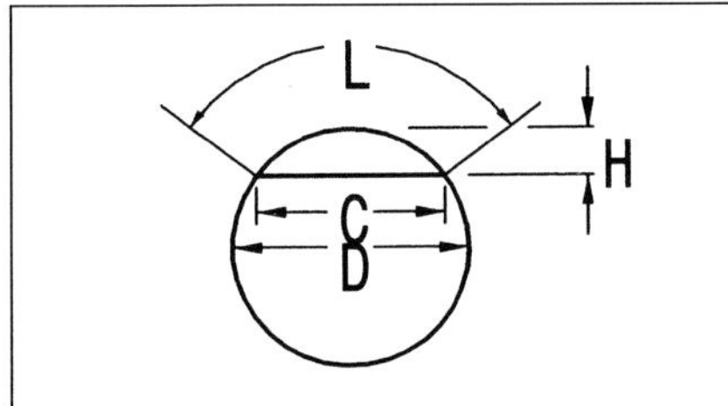

Arc length (L) = $0.00873 \times D \times La$
where:

La = angle of arc

Chord (C) = $2\sqrt{H(D - H)}$

$$\text{Height (H)} = \frac{D}{2} - \frac{1}{2}\sqrt{D^2 - C^2}$$

SOLIDS

CYLINDER:

$$\text{Volume} = \frac{\pi \times D^2}{4} \times L$$

SPHERE:

$$\text{Volume} = \frac{\pi \times D^3}{6}$$

CONE:

$$\text{Volume} = \frac{\pi \times D^2 \times H}{12}$$

where:

D = diameter

L = length of cylinder

H = height of cone

RECTANGLE

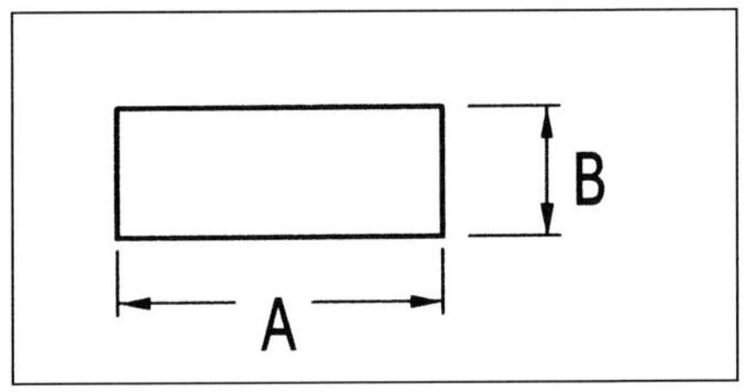

Area = $A \times B$

TRIANGLE

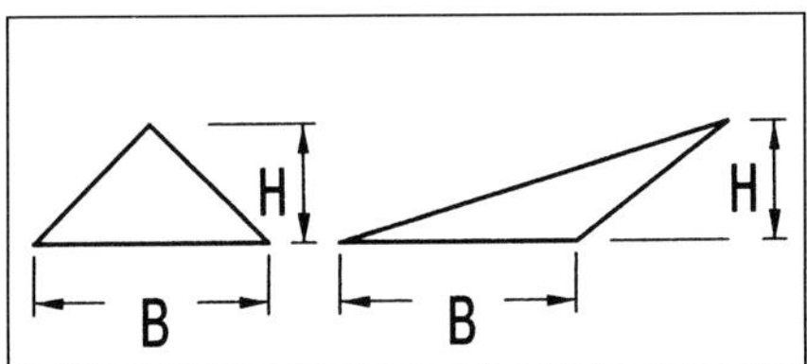

$$\text{Area} = \frac{B}{2} \times H$$

Length of the third side (A) of a right-angle triangle

if the lengths of the other two sides (B and C) are known (Pythagoras's theorem):

$$A = \sqrt{B^2 + C^2}$$

TRAPEZOID

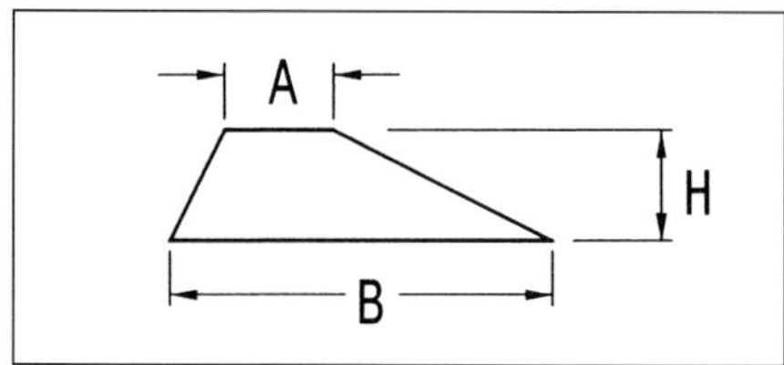

$$\text{Area} = \frac{A + B}{2} \times H$$

HEXAGON

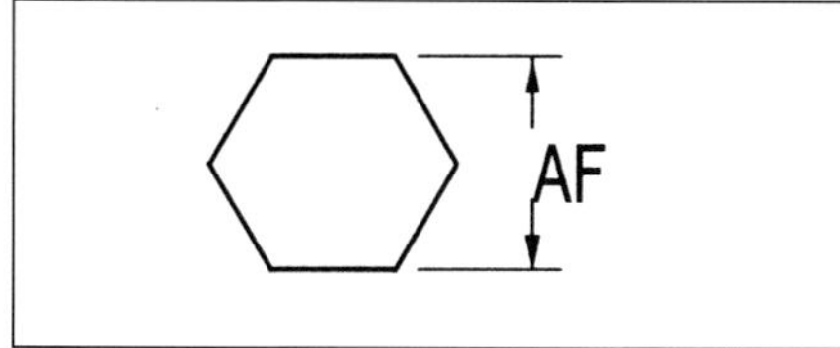

$$\text{Area} = 0.866 \times AF^2$$

where:

AF = across flat

RELATIONSHIP BETWEEN SHAPES AND CONTAINING CIRCLES

EQUILATERAL TRIANGLE:

$$D = 1.155 \times A$$

SQUARE:

$$D = 1.415 \times A$$

HEXAGON:

$$D = 1.155 \times A$$

OCTAGON:

$$D = 1.083 \times A$$

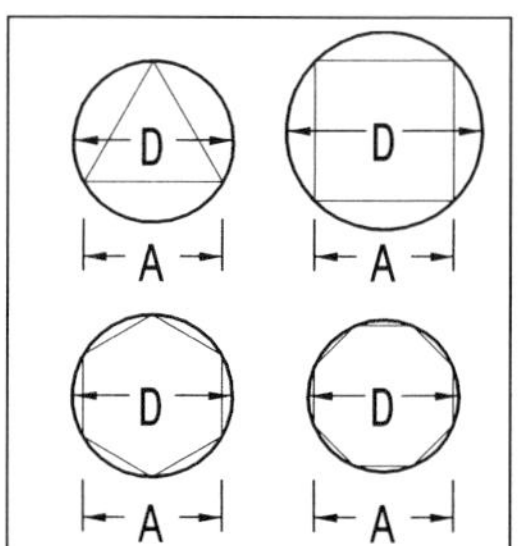

In view of the probable uses for the above in the workshop, all the constants are rounded up.

SETTING OUT A RIGHT ANGLE

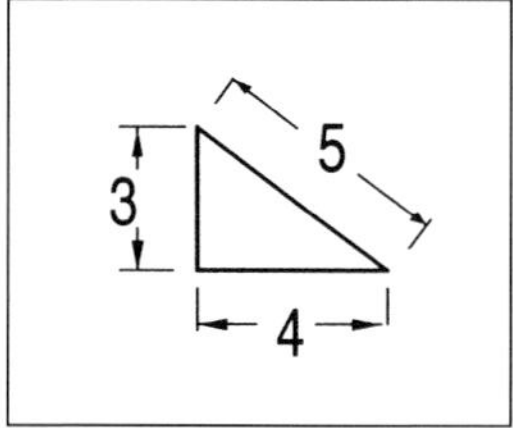

A triangle with sides in the proportions of 3, 4 and 5 will produce an exact right angle.

TRIGONOMETRICAL FORMULAS

These are used to work out the values of unknown sides and angles of triangles.

FOR RIGHT-ANGLE TRIANGLES ONLY

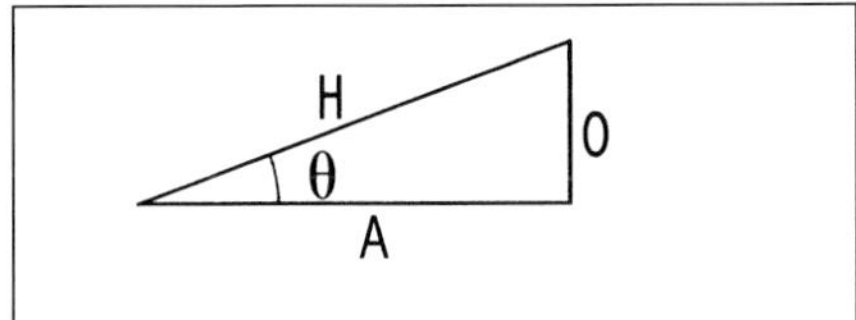

Only two values need to be known, of which at least one must be a length.

$$\sin \theta = \frac{O}{H}$$

$$\cos \theta = \frac{A}{H}$$

$$\tan \theta = \frac{O}{A}$$

where:

θ = angle
O = opposite
H = hypotenuse
A = adjacent

FOR ANY TRIANGLE

In all triangles, the sum of the three angles always equals 180°:

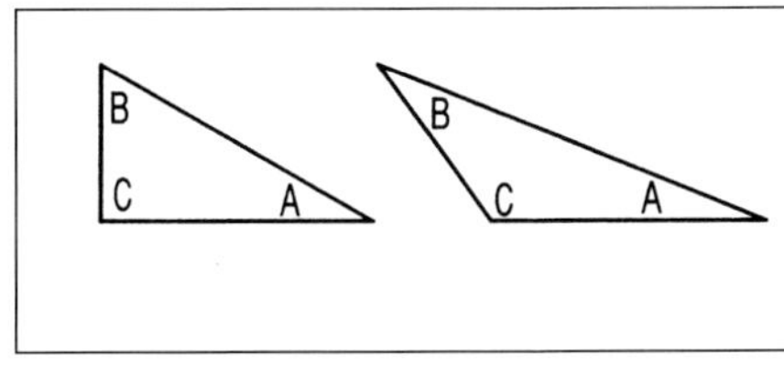

$$A + B + C = 180°$$

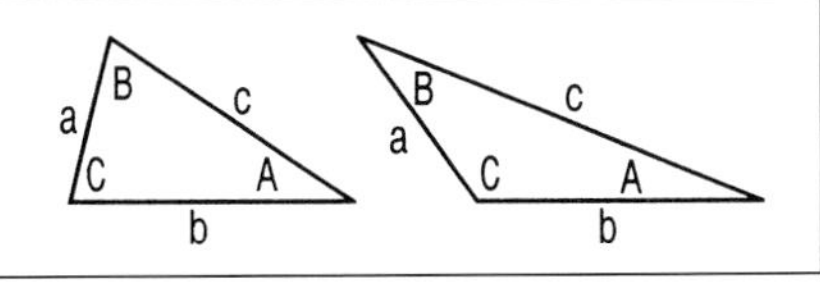

$$\frac{a}{\sin A} = \frac{b}{\sin B} = \frac{c}{\sin C}$$

If three values (one length and two angles) are known, the other lengths can be calculated using the above, as follows:

$$a = \frac{b \times \sin A}{\sin B} \quad \text{or} \quad = \frac{c \times \sin A}{\sin C}$$

$$b = \frac{c \times \sin B}{\sin C} \quad \text{or} \quad = \frac{a \times \sin B}{\sin A}$$

$$c = \frac{a \times \sin C}{\sin A} \quad \text{or} \quad = \frac{b \times \sin C}{\sin B}$$

If three values (two lengths and either of the angles opposite one of the known lengths) are known, the other angles can be calculated as follows:

$$\sin A = \frac{a \times \sin B}{b} \quad \text{or} \quad = \frac{a \times \sin C}{c}$$

$$\sin B = \frac{b \times \sin C}{c} \quad \text{or} \quad = \frac{b \times \sin A}{a}$$

$$\sin C = \frac{c \times \sin A}{a} \quad \text{or} \quad = \frac{c \times \sin B}{b}$$

If three values (two lengths and the angle between them) are known, the other angles and the unknown side can be calculated as follows.

For sides a and b and angle C:

$$\tan A = \frac{a \times \sin C}{b - (a \times \cos C)}$$

$$B = 180 - (A + C)$$

$$c = \frac{a \times \sin C}{\sin A}$$

For sides b and c and angle A:

$$\tan B = \frac{b \times \sin A}{c - (b \times \cos A)}$$

$$C = 180 - (B + A)$$

$$a = \frac{b \times \sin A}{\sin B}$$

For sides c and a and angle B:

$$\tan C = \frac{c \times \sin B}{a - (c \times \cos B)}$$

$$A = 180 - (C + B)$$

$$b = \frac{c \times \sin B}{\sin C}$$

If only three lengths are known, the angles can be calculated as follows:

$$\cos A = \frac{b^2 + c^2 - a^2}{2bc}$$

$$\cos B = \frac{a^2 + c^2 - b^2}{2ac}$$

$$\cos C = \frac{a^2 + b^2 - c^2}{2ab}$$

CHAPTER 18
T-SLOTS AND DOVETAILS

T-SLOTS AND T-NUTS

When designing equipment that requires the inclusion of T-slots it is best to work to standard dimensions. This will be inevitable if machining with a standard T-slot cutter, provided that it has not been repeatedly resharpened, thus making it below the recommended minimum size.

STANDARD METRIC SIZES

Slots and cutters

Table 18.1 gives the standard dimensions for sizes M4 to M16 and the maximum and minimum values for dimensions A1, T1 and H1 (see drawing). Those for A1 and T1 determine the dimensions for the T-slot cutter and are the maximum to which a new cutter should conform. The minimum dimension should enable the cutter to be resharpened a number of times, thus delaying the time for a replacement to be required.

Dimension H1 should, for strength, ideally be made to the larger dimension. The wide tolerance permits the value to be reduced where compactness of design is desirable.

S1 is a nominal dimension but should be made with a plus tolerance, e.g. +0.2 or +0.3, to avoid it being a tight fit on the T nuts used with it.

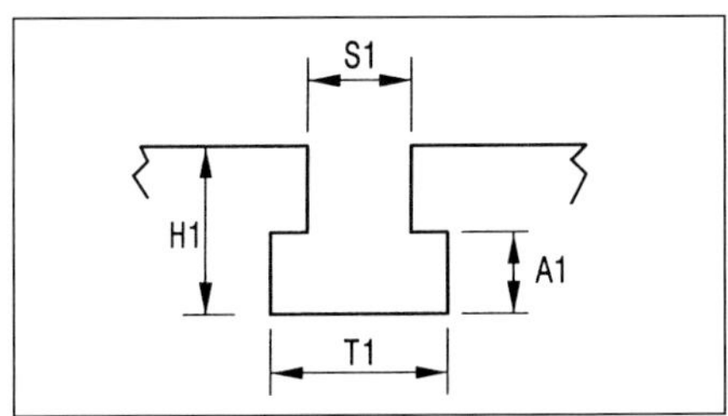

Table 18.1 Standard metric sizes of T-slot (mm)

Size	S1	T1	A1	H1
M4	5	10–11	3.5–4.5	8–10
M5	6	11–12.5	5–6	11–13
M6	8	14.5/16	7–8	15–18
M8	10	16–18	7–8	17–21
M10	12	19–21	8–9	20–25
M12	14	23–25	9–11	23–28
M16	18	30–32	12–14	30–36

Nuts

If making T-nuts, do not take the dimensions from the slots in the machine on which they are originally intended to be used but work to the values in Table 18.2. This will ensure that the nuts will be usable, even with a machine that has slots to the minimum dimensions.

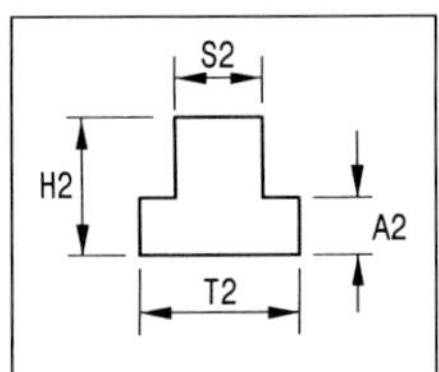

Size	S2	T2	A2	H2
M4	5	9	2.5	6
M5	6	10	4	8
M6	8	13	6	10
M8	10	15	6	12
M10	12	18	7	16
M12	14	22	8	19
M16	18	28	10	25

Dimension S2 (see drawing) is a nominal dimension but should be made with a minus tolerance, e.g. −0.2 or −0.3, to avoid it being a tight fit in the T-slots in which it is used.

The length of the nut should not be less than its maximum width.

Studs

Studs with heads are made to the same dimensions as the T nuts as far as these are appropriate.

Drilling machines

While smaller drilling machines do not have T-slotted tables the open bottom slots are often made to a dimension to accept standard T nuts; this may not be apparent.

MACHINE TOOL SLIDE DOVETAILS

There are no standards for the dimensions of dovetail slides, other than the angle, which is governed by the cutters available, usually 45° and 60°. Of course, if the slide is machined on a shaper then any angle can be achieved. This method has a lot to recommend it if a shaper is available.

DESIGN

Because no standard exists, the designer has a choice of some dimensions, from which the remaining dimensions need to be calculated, enabling machining to commence. These are shown in the series of drawings (D1–D7).

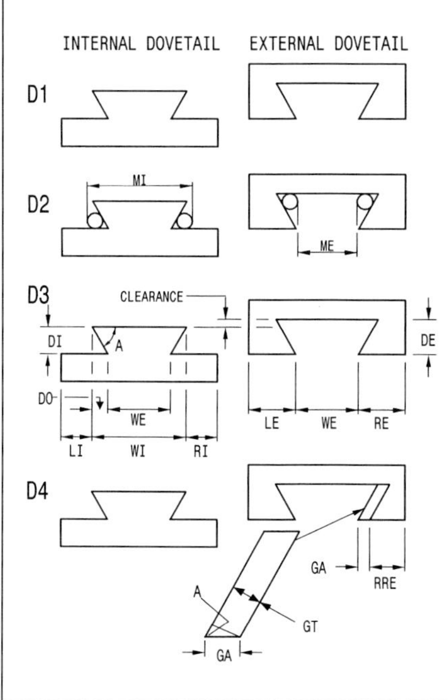

Frequently the dimension for the final width (internal and external dovetails) is given as the width across two rods placed into the dovetail groove(shown as MI and ME in D2).

However, before arriving at this point, the machining will consist of many stages, each of which require working to dimensions in addition to this single dimension given.

Design dimensions

In the drawing, D3 shows the dimensions required at the initial stage, these fall into two groups: those that can be chosen by the designer and those that have to be calculated from the chosen dimensions.

Dimensions that can be chosen:

WI   Maximum width of internal dovetail

LI   Left-hand edge to dovetail edge, internal

RI   Right-hand edge to dovetail edge, internal

DI   Depth of dovetail, internal

DE   Depth of dovetail, external

A   Dovetail angle.

In many cases LI and RI will be equal, but are referred to separately as this is not always be the case.

Dimensions that need to be calculated:

DO   Dovetail overhang, internal.

WE   Minimum width of internal dovetail

LE   Left-hand edge to dovetail edge, external

RE   Right-hand edge to dovetail edge, external.

Once the dimensions in the first group have been chosen, the dimensions in the second group can be calculated as follows:

$$DO = \frac{DI}{\tan A}$$

$$WE = WI - 2DO$$

$$LE = LI + DO$$

$$RE = RI + DO$$

Gib strip

A gib strip will almost always be fitted and one side of the external dovetail will have to be reduced to make space for it (see D4)

Dimension that can be chosen:

GT   Gib strip thickness.

Dimensions that need to be calculated:

GA   Gib strip allowance

RRE   Reduced right-hand edge to dovetail edge, external.

$$GA = \frac{GT}{\sin A}$$

$$RRE = RE - GA$$

Note

To make sure that there is enough clearance to permit the gib strip to be fitted, use a value of GA that is slightly greater than the calculated value, say 3.2mm for a calculated value of 3mm.

Test dimensions

After completing the initial dimensions, there now remain the dimensions for checking the dovetails (as in D2 and D5).

Dimension that can be chosen:

R   Diameter of test rod, R being its radius.

Dimensions that need to be calculated:

N   Center of testing rod to corner of dovetail

DOE   Dovetail overhang, external

MI   Check dimension for internal dovetail

ME   Check dimension for external
      dovetail.

$$N = \frac{R}{\operatorname{Tan} \dfrac{A}{2}}$$

$$MI = WE + 2\,(N + R)$$

$$DOE = \frac{DE}{\tan A}$$

$$ME = WE + 2\,(DOE - N - R)$$

Important note

Note that the dimension ME for the external dovetail is based on no gib strip being fitted. In most cases this will need to be increased by the value GA (see D4) to allow space for one.

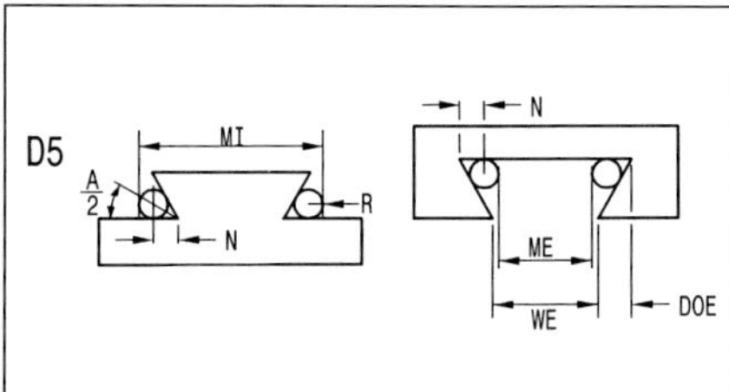

MANUFACTURE

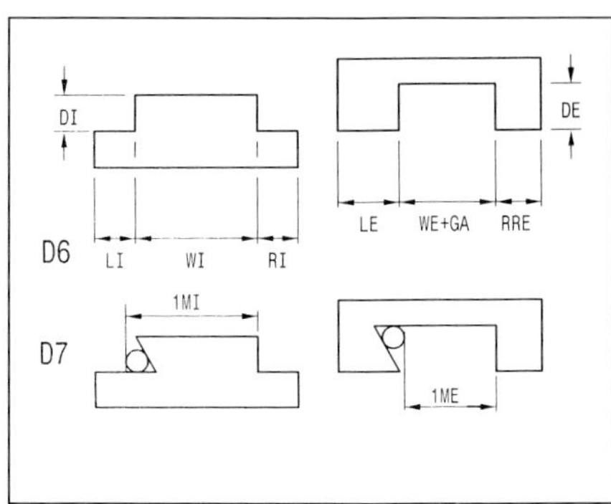

This section assumes that a gib strip is fitted because this is the most common arrangement.

The first stage is to machine with a rectangular form (as shown in D6) all the dimensions from those already chosen or calculated. Note that, while DE is clearance only over DI, it must be to dimension, as any variation will change the value required for ME and 1ME.

Next the first side of the dovetail has to be machined (as shown in D7). This requires values for 1MI and 1ME to be established.

$$1MI = WI - DO + N + R$$
$$1ME = WE + GA + DOE - N - R$$

Finally the second side of each dovetail is machined to dimensions MI and RME.

$$RME = ME + GA$$

CHAPTER 19
ELECTRICAL/ELECTRONIC COMPONENTS

ELECTRONIC COMPONENT MARKING

RESISTORS

Small resistors, intended for electronic circuits, have their values indicated by a series of colored bands. If there are four bands, the first two indicate the first two digits while the third indicates the multiplier, e.g. forYellow-Violet-Orange would be 47000 ohms. The fourth band indicates the tolerance (+ and -).

Standard values

Resistor values are made to a number of standard series. Four-band resistors usually conform to series E6, E12 or E24, 6, 12, and 24 being the number of values between 1 and 9.9, 10 and 99, 100 and 990, and so on (see Table 19.1).

Table 19.1 Standard values of resistors

E6	10	-	-	-	15	-	-	-
E12	10	-	12	-	15	-	18	-
E24	10	11	12	13	15	16	18	20
E6	22	-	-	-	33	-	-	-
E12	22	-	27	-	33	-	39	-
E24	22	24	27	30	33	36	39	43
E6	47	-	-	-	68	-	-	-
E12	47	-	56	-	68	-	83	-
E24	47	51	56	62	68	75	82	91

Typically therefore, the E6 series has values of 1.0, 1.5, 2.2, 3.3, 4.7, 6.8, 10, 15, 22, 33, 47, 68, 100, etc.

Series E48 and E96 have smaller divisions, the E96 series starting at 100, 102, 105, 107 and 110. As they have three digits, an extra band is required. Thus the first three bands indicating the three digits, the fourth the multiplier, and the fifth the tolerance. In some cases a sixth band indicates the temperature coefficient.

Color codes

Digits:

Black	0
Brown	1
Red	2
Orange	3
Yellow	4
Green	5
Blue	6
Violet	7
Gray	8
White	9

Multipliers:

Black	× 1 (add no zero)
Brown	× 10 (add 1 zero)
Red	× 100 (add 2 zeros)
Orange	× 1 000 (add 3 zeros)
Yellow	× 10 000 (add 4 zeros)
Green	× 100 000 (add 5 zeros)
Blue	× 1 000 000 (add 6 zeros)
Violet	× 10 000 000 (add 7 zeros)

Tolerances:

Brown	±1%
Red	±2%
Gold	±5%
Silver	±10%

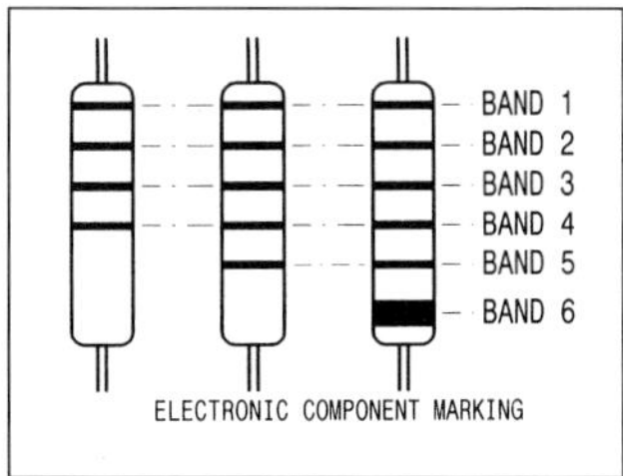

Resistors marked with their values

In some cases, resistors are marked directly with their values, quoted in ohms or, for the higher values, kilohms and megohms. In these values, the letter R is used to indicate ohms, K to indicate kilohms and M to indicate megohms. To avoid any misreading of the decimal point, the multipliers R, K and M are positioned in place of the decimal point, as in the following examples:

R33	0.33 ohms
1R5	1.5 ohms
33R	33.0 ohms
2K2	2.2 kilohms
68K	68.0 kilohms
4M7	4.7 megohms

For the same reason, this method is also used on drawings.

CAPACITORS

Capacitor marking is much more varied and, as a result, cannot be covered fully. However, the following information may be of some help in certain situations, but it needs to be treated with some caution. On physically larger capacitors, the value and voltage rating, and possibly the tolerance, will be printed in full, but this is not possible on smaller capacitors.

Standard values

Capacitor values also conform to the standard E series quoted for resistors (see previous) but very specific values are rarely needed. As a result, fewer values are available, conforming only to E3, E6 or E12. For example, the values for E3 (in picofarads) are: 100pf, 220pf, 470pf, 1000pf, 2200pf, 4700pf, and so on.

Marking methods

As well as being printed on the capacitor, the values can also be conveyed by means of a color or an alphanumeric code. The alphanumeric code is quite common and consists of three numbers, which are a code for the value, and a letter for the tolerance. The first two numbers are the first two digits of the value, while the third is the number of added zeros, giving the value in picofarads. For example, 473J is 47000pf (being 0.047µf), and J indicates 5% tolerance.

Tolerance letter:

F	±1%
G	±2%
H	±2.5%
J	±5%
K	±10%
M	20%

Additional characters are added to indicate the manufacturer's type code, but no explanation of these can be given.

Color-coded capacitors, although rare, follow a similar approach, with the first two colors representing the first two digits and the third the number of added zeros; again the value is in picofarads. Additional colors indicate the tolerance and voltage rating. The color code for the value is the same as that quoted for resistors (see above), but the color code for the tolerance may differ.

ELECTRIC MOTORS

Electric motors can be divided into those powered by AC and those powered by DC, although a few commutator motors can be used on either supply.

CHARACTERISTICS

Reversing

In theory, most motors are reversible, but the connections may be permanently made inside the motor for a single direction. Most often motors run in one direction only, but it may be necessary to reverse the direction initially in order to give the direction required by the machine being driven; this is done by altering the connections to the motor. If a motor is to run in either direction, switches and/or a reversing starter will be required and, in most cases, it is highly desirable, if not essential, for the motor to return to the resting state before starting again in reverse. Because of this, providing the necessary delay before a change in direction should be inherent in the controls rather than being dependent on the operator to provide the necessary delay between one direction and the other. Therefore, you should ensure that you understand the requirements and, if in doubt, refer to a specialist book on the subject.

Speed control

The speed of most motors can be varied and, in some cases, depending on the level of speed holding required, with quite simple equipment. In other cases the equipment is very complex.

Speed holding

The ability to hold speed with changes in load depends on the type of motor. Details given are for the motor only. The supply voltage, fixed or from a simple speed controller, may also vary with load, thereby increasing the speed variation. More complex controllers can detect changes and compensate, achieving a much better speed holding than can be obtained by the motor alone.

AC 3-PHASE INDUCTION MOTOR

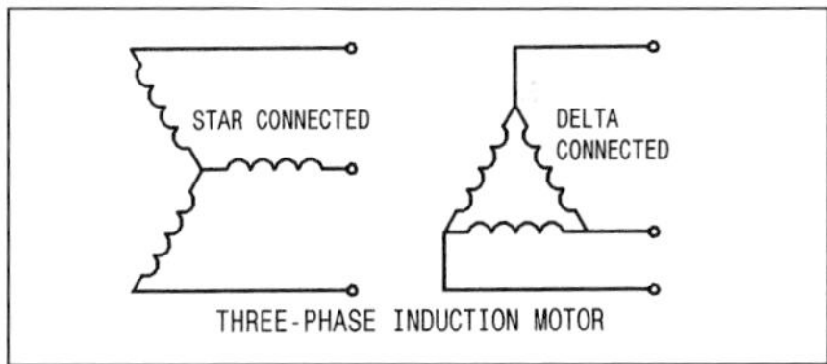

Reversing: Interchange the connections to any two of the three terminals.

Speed control: Only by varying the supply frequency using complex electronic equipment.

Speed holding: Very good. Speed controllers also have good speed holding characteristics.

Supply voltage: Modern 3-phase motors have six terminals, enabling them to be star or delta connected. The voltage required in star will be 1.73 times that in delta. Small motors will mostly be 380/420V (star) and 220/250V (delta). Many speed controllers for small 3-phase motors need a 200/240V single-phase input, giving a 220V 3-phase output, and require the motor to be delta connected.

AC SINGLE-PHASE INDUCTION MOTORS

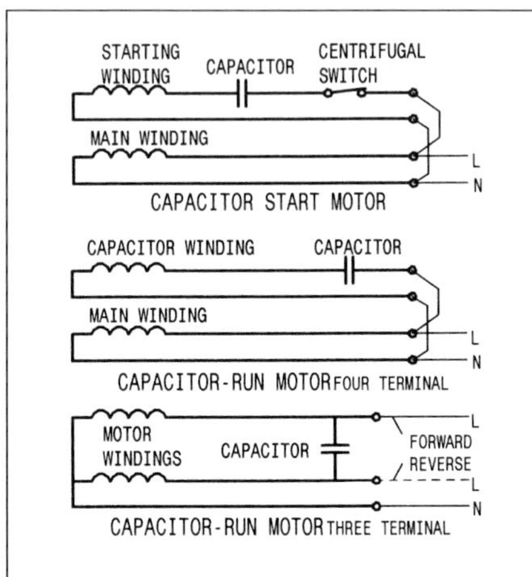

Reversing: For a capacitor start and four-terminal capacitor-run motors, interchange the connections to one or other of the windings, but not to both. Three-terminal capacitor-run motors are reversed by connecting the capacitor in series with one or other of the windings.

Speed control: The speed of a capacitor-run motor can be varied by varying the supply frequency. This uses complex electronic equipment and the motor capacitor will probably have to be removed. In the case of capacitor start motors, the centrifugal switch will close as the speed is reduced, reconnecting the starter winding; this is not acceptable.

Speed holding: Very good.

AC SINGLE-PHASE COMMUTATOR MOTORS

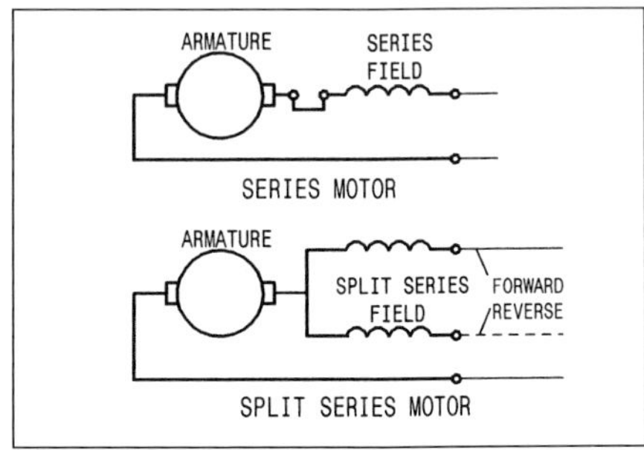

These motors are always series-connected, hand-power tools being a common application. Series motors intended for use on AC will also run on DC. Those specifically designed with this in mind will state this on the motor nameplate. However, series motors designed to run on a DC supply should not be run on an AC supply.

Reversing: Interchange the connections to the armature or the field, but not to both. Split series motors are reversed by connecting to one or other of the windings only.

Speed control: This can be varied by varying the supply voltage.

Speed holding: Poor

DC SHUNT-WOUND MOTORS

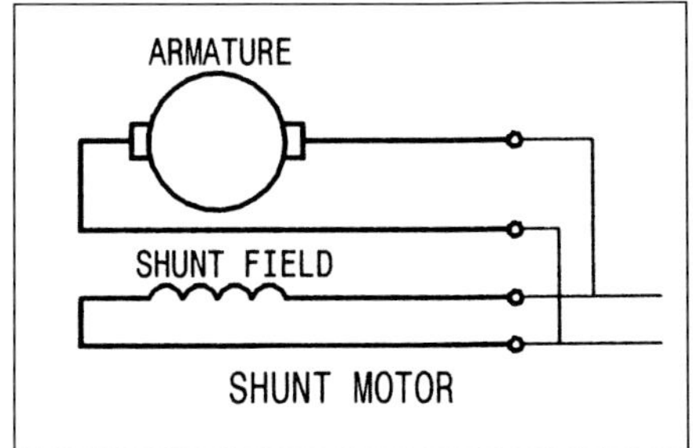

Reversing: Interchange the connections to the armature or the shunt field, but not to both.

Speed control: Possible by varying the voltage but special conditions apply. Both armature and field can be varied, but not together because reducing armature voltage reduces speed while reducing field voltage increases speed.

To reduce the speed below normal, the normal practice is to reduce the armature voltage while holding the field voltage at the maximum value. To increase the speed above the normal, the field voltage is reduced while the armature voltage is held at maximum value.

For the widest speed range, both methods can be used, but not simultaneously. Increasing speed by reducing the field voltage must take into account the maximum permitted speed of the motor because the motor can be run at considerably above this speed if the field voltage is reduced substantially, particularly if it is lightly loaded. In the absence of a definite figure, an increase of 50% should be considered the maximum.

Speed will vary approximately in direct proportion to the armature voltage and in inverse proportion to the field voltage

Speed holding: Good.

PERMANENT MAGNET MOTORS

These have similar characteristics to the shunt motor but the magnetic field is provided by a permanent magnet rather than the shunt winding.

Reversing: Reverse the supply to the two armature terminals.

Speed control: Possible by varying the supply voltage to the armature.

Speed holding: Good.

DC SERIES-WOUND MOTORS

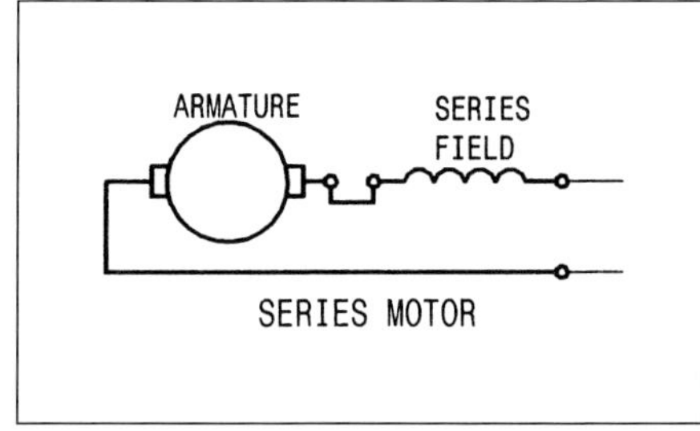

Reversing: Interchange the connections to the armature or the field, but not to both.

Speed control: Possible by varying the supply voltage.

Speed holding: Poor.

DC MOTOR STARTING

The current taken by a DC motor while it is running is mainly limited by its own internally generated opposing voltage, or back electromotive force (BEMF). As this is proportional to speed, the BEMF is zero when the motor is stationary and the current is limited only by the motor's resistance. This is a particular problem with a shunt-wound motor, where the supply is connected directly to the armature. This may have a low resistance, resulting in a large starting current, which can be a problem, particularly with larger motors and those intended for 110V or 220V supplies. In these cases, some means of applying the voltage gradually, either by voltage control or series resistors, is essential. This is rarely a problem with very small motors but it is useful to aware of this requirement. Even with small motors, applying the voltage in two stages using a series resistor, switched out after starting, may prove beneficial.

Motor on/off controls

Smaller motors can be switched using manual switches, but take note of their required current and voltage ratings, especially if mains fed.

If switching the field of a shunt motor for reversing, be sure to use a switch rated for DC because inductive DC circuits can be a problem.

Induction motors, single or three-phase, should be switched with a starter that has overload protection. This is a condition of the motor manufacturer's guarantee.

Direct on line (DOL) starters

A DOL starter is one that applies the supply to the motor in one step, with no means of increasing the voltage gradually.

No volt release (NVR) starters

This is the desired method of control and is mandatory in industry for many applications, e.g. machine tools. These starters are fitted with on/off push buttons and a coil-operated contactor, which ensures that, in the event of loss of power for any reason, the motor will not restart when the power is restored unless the start button is pressed again.

They also permit the addition of remote controls, typically a foot switch on a drilling machine. Note that, in some controllers, the on/off buttons are just rocker switches and do not possess the NVR feature.

Overload protection

NVR starters are normally fitted with an overload protection device but the current range covered by this is limited so the starter must be chosen to suit the full load current of the motor it is to control. Modern mains motors are designed to run at a high temperature, very close to the maximum permitted, and because of this the overload must not be set above the current given on the motor nameplate. If the overload frequently trips, do not increase its setting; this means that the motor is too small for the application.

CHAPTER 20
ELECTRICAL FORMULAS

RESISTORS IN SERIES

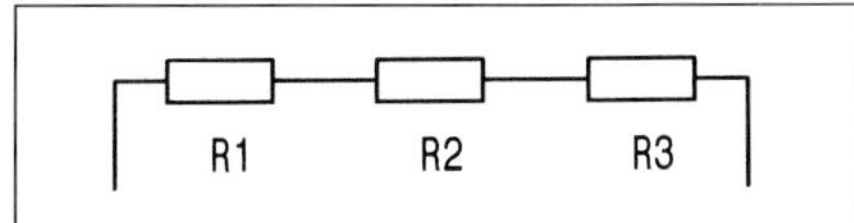

$$R = R1 + R2 + R3$$

RESISTORS IN PARALLEL

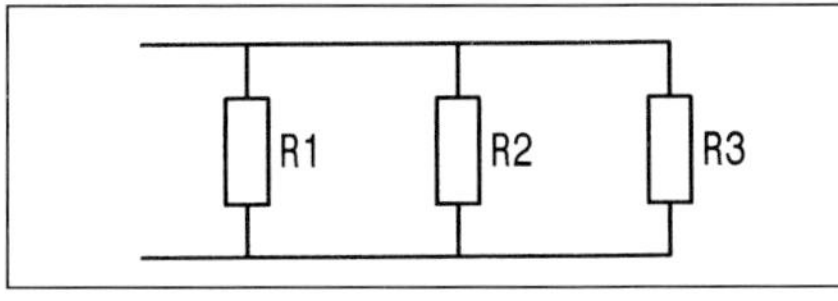

$$R = \cfrac{1}{\cfrac{1}{R1} + \cfrac{1}{R2} + \cfrac{1}{R3}}$$

CAPACITORS IN SERIES

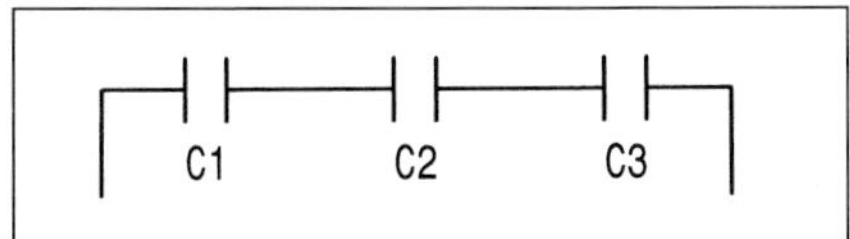

$$C = \cfrac{1}{\cfrac{1}{C1} + \cfrac{1}{C2} + \cfrac{1}{C3}}$$

CAPACITORS IN PARALLEL

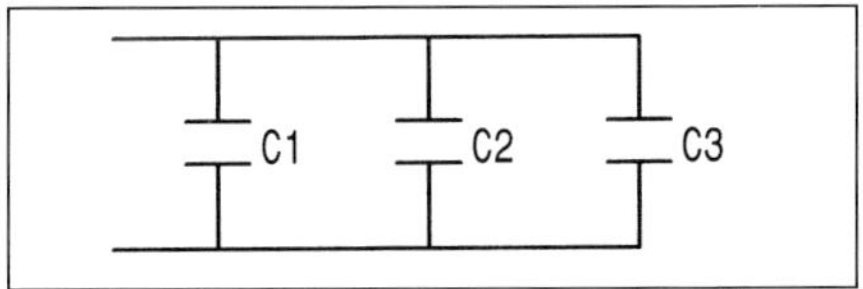

$$C = C1 + C2 + C3$$

In the above, R (resistance) and C (capacitance) are the effective value of the series or parallel configuration. Inductors follow the same principle as resistors.

RELATIONSHIP BETWEEN VOLTAGE, RESISTANCE AND CURRENT (OHM'S LAW)

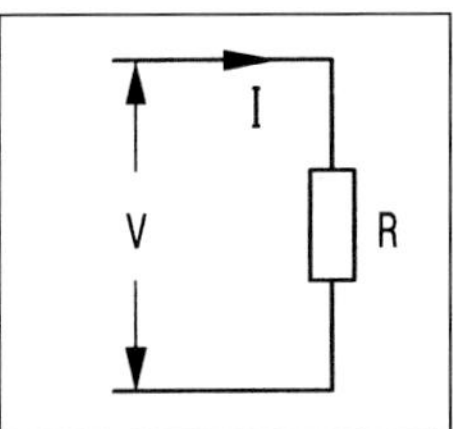

$$I = \frac{V}{R}$$

$$V = I \times R$$

$$R = \frac{V}{I}$$

where:

R = resistance in ohms

V = voltage in volts

I = current in amps.

Relationship between current, voltage and power:

$$P = V \times I$$

Also, in terms of resistance:

$$P = \frac{V^2}{R}$$

$$P = I^2 \times R$$

where:

P = power in watts.

The above are applicable to DC (direct current) circuits, but can also be applied to AC (alternating current) circuits, provided that the load is pure resistance, i.e. there is no inductive or capacitive content. Heater elements, even if wound, normally have a high resistive to inductive ratio and the above can be used with negligible error. The formulas do not apply to AC circuits with inductive loads, such as motors, solenoids and transformer windings.

Ohm's law is applicable not only to a single resistor applied to a single power source, but also to individual elements of a complex circuit. The following examples give some indication of its use in these situations. The first two examples indicate the principles of the addition of voltages and currents in multi-element circuits.

VOLTAGES AND CURRENTS IN COMPLEX CIRCUITS

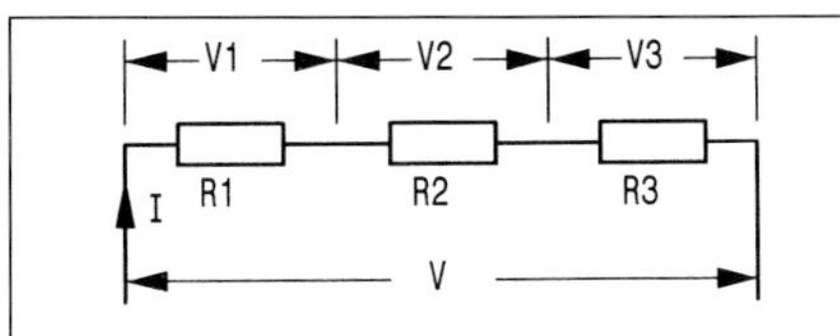

$$V = V1 + V2 + V3$$

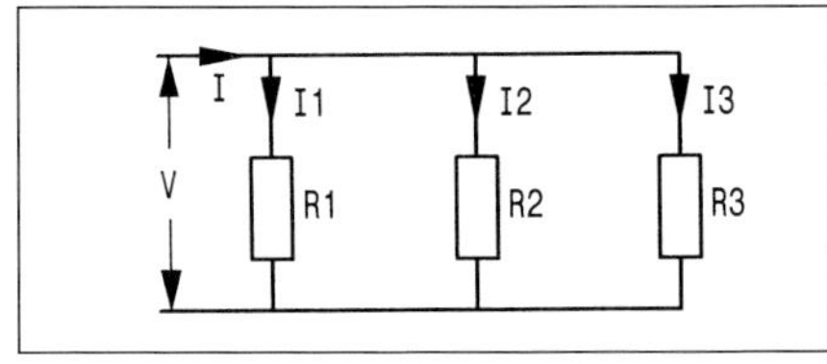

$$I = I1 + I2 + I3$$

Applying Ohm's Law (V = R × I) to individual elements of a series circuit:

$$V1 = R1 \times I$$

$$V2 = R2 \times I$$

$$V3 = R3 \times I$$

Applying Ohm's Law (I = V/R) to individual elements of a parallel circuit:

$$I1 = \frac{V}{R1}$$

$$I2 = \frac{V}{R2}$$

$$I3 = \frac{V}{R3}$$

While all these examples given relate to three resistors, capacitors, voltages, etc., the formulas suit any number from two or more.

SERIES/PARALLEL CIRCUITS

There are more complex circuits, other than purely series or purely parallel, and voltages and currents in these can also be calculated using the formulas. A typical example is the use of a potentiometer to supply a load (see drawing). This consists of a series/parallel circuit and, while it is only made up of two components, for calculation purposes it can be considered as being made up of three resistors.

To calculate voltages V1 and V2 in the theoretical circuit (also shown in the drawing), it is necessary to work out the effective values of R2 and R3, using the 'two resistors in parallel' formula. Then, using this value, calculate V1 and V2 as though there were just two resistors in series.

AC CIRCUITS

In AC circuits that are other than purely resistive, the term 'resistance' is replaced by the term 'impedance'. Impedance is the total effect of the resistive, inductive and capacitive elements of the circuit. These elements may be found in isolation or in any combination and are also frequency dependent. This subject is too complex to cover here and readers wishing to make calculations relating to AC circuits are advised to consult more detailed information.

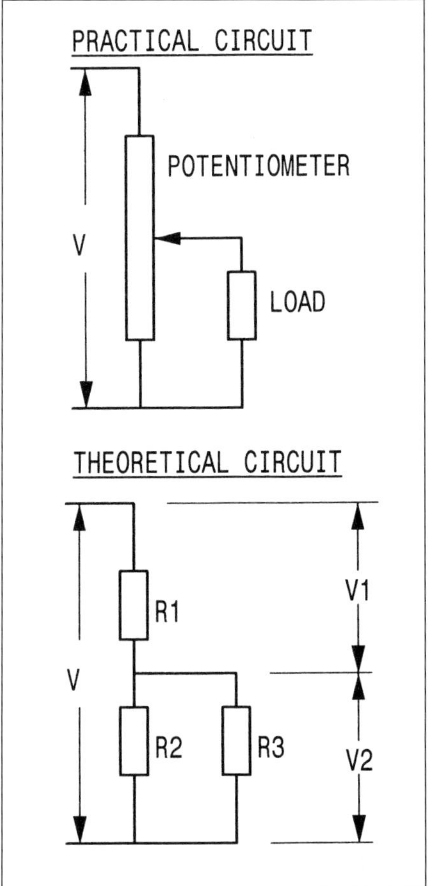

Potentiometer + theoretical circuit

CHAPTER 21

HARDWARE DIMENSIONS

STANDARD SCREW DIMENSIONS

SOCKET-HEAD CAP SCREWS: ISO METRIC MICRO SCREWS

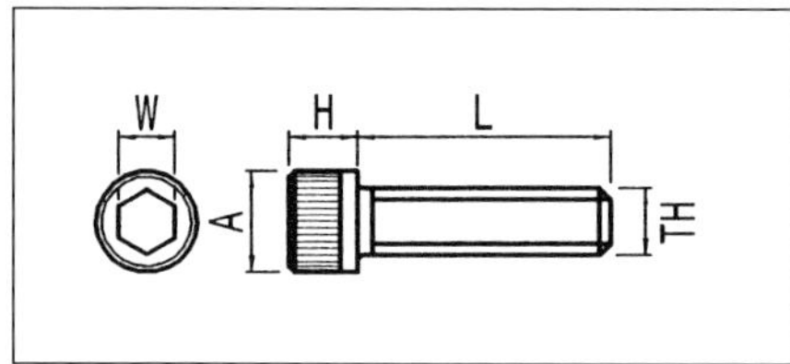

Table 21.1. ISO metric micro screws: sizes and dimensions (mm)

TH	A	H	L(1)	W
M1.4	2.6	1.4	6	1.27
M1.6	3.0	1.6	6	1.5
M1.8	3.4	1.8	6	1.5
M2.0	3.8	2.0	12	1.5
M2.5	4.5	2.5	15	2.0

SOCKET-HEAD CAP SCREWS: METRIC AND IMPERIAL

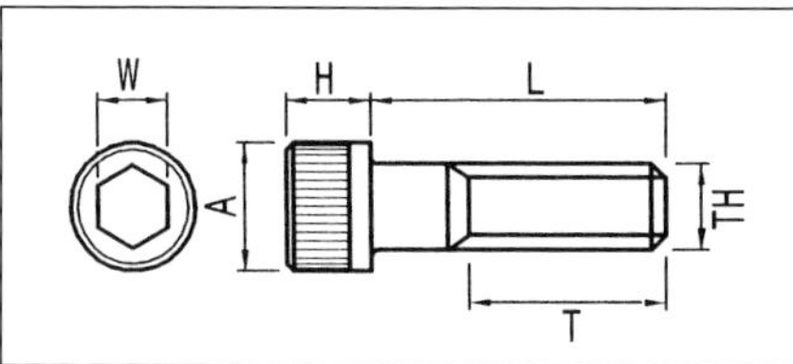

Table 21.2. ISO metric cap screws: sizes and dimensions (mm)

TH	A	H	L(2)	W	T
M3	5.5	3	20	2.5	18
M4	7.0	4	25	3.0	20
M5	8.5	5	25	4.0	22
M6	10.0	6	30	5.0	24
M8	13.0	8	35	6.0	28
M10	16.0	10	40	8.0	32
M12	18.0	12	50	10.0	36
M16	24.0	16	60	14.0	44
M20	30.0	20	70	17.0	52

Table 21.3. BA cap screws: sizes and dimensions (in)

TH	A	H	L(2)	W	T(3)
8BA	0.140	0.087	3/4	1/16	5/8
7BA	0.161	0.098	3/4	1/16	5/8
6BA	0.187	0.110	7/8	5/64	3/4
5BA	0.219	0.126	7/8	3/32	3/4
4BA	0.219	0.142	1	3/32	3/4
3BA	0.250	0.161	1	1/8	7/8
2BA	0.312	0.187	1	5/32	7/8
1BA	0.312	0.209	1 1/4	5/32	1
0BA	0.375	0.236	1 1/4	3/16	1

Table 21.4. UNF/UNC cap screws: sizes and dimensions (in)

TH	A	H	L(2)	W	T(3)
1	0.118	0.073	3/4	1/16	5/8
2	0.140	0.086	3/4	5/64	5/8
3	0.161	0.099	3/4	5/64	5/8
4	0.183	0.112	7/8	3/32	3/4
5	0.205	0.125	7/8	3/32	3/4
6	0.226	0.138	1	7/64	3/4

8	0.270	0.164	1	9/64	7/8
10	0.312	0.190	1	5/32	7/8
12	0.343	0.216	1	5/32	7/8
1/4	0.375	0.250	1 1/4	3/16	1
5/16	0.468	0.312	1 1/2	1/4	1 1/8
3/8	0.562	0.375	1 1/2	5/16	1 1/4
7/16	0.656	0.437	1 3/4	3/8	1 3/8
1/2	0.750	0.500	2	3/8	1 1/2
9/16	0.843	0.562	2 1/4	7/16	1 5/8

TH	A(4)	A1	H	L(2)	W	T
1BA	0.396	0.360	0.093	1	3/32	1
0BA	0.449	0.408	0.106	1	1/8	1 1 1/2

Countersunk angle = 90°

Table 21.7. UNF/UNC countersunk screws: sizes and dimensions (in)

TH	A(4)	A1	H	L(2)	W	T(3)
1	0.168	0.143	0.054	3/4	0.05	5/8
2	0.197	0.168	0.064	3/4	0.05	5/8
3	0.226	0.193	0.073	3/4	1/16	5/8
4	0.255	0.218	0.083	1	1/16	3/4
5	0.281	0.240	0.090	1	5/64	3/4
6	0.307	0.263	0.097	1	5/64	3/4
8	0.359	0.311	0.112	1	3/32	7/8
10	0.411	0.359	0.127	1 1/4	1/8	7/8
1/4	0.531	0.480	0.161	1 1/2	5/32	1
5/16	0.656	0.600	0.198	1 3/4	3/16	1 1/8
3/8	0.781	0.720	0.234	2	7/32	1 1/4
7/16	0.844	0.781	0.234	2 1/4	1/4	1 3/8

Countersunk angle = 80°

SOCKET-HEAD COUNTERSUNK SCREWS

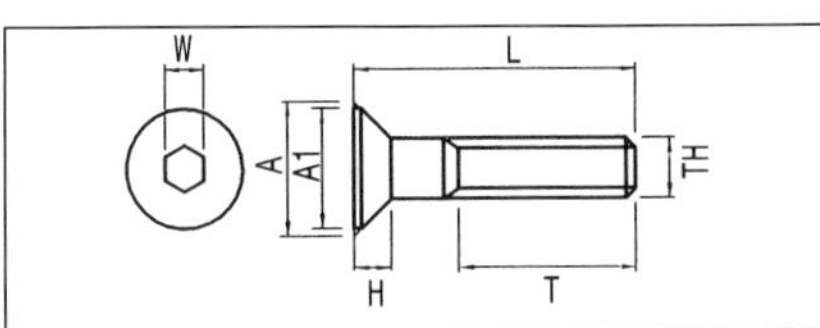

Table 21.5. ISO metric countersunk screws: sizes and dimensions (mm)

TH	A (4)	A1	H	L(2)	W	T
M3	6.72	5.82	1.86	30	2.0	18
M4	8.96	7.78	2.48	30	2.5	20
M5	11.20	9.78	3.10	35	3.0	22
M6	13.44	11.73	3.72	40	4.0	24
M8	17.92	15.73	4.96	45	5.0	28
M10	22.40	19.67	6.20	55	6.0	32
M12	26.88	23.67	7.44	60	8.0	36
M16	33.60	29.67	8.80	70	10.0	44
M20	40.32	35.61	10.16	100	12.0	52

Countersunk angle = 90°

Table 21.6. BA countersunk screws: sizes and dimensions (in)

TH	A(4)	A1	H	L(2)	W	T
8BA	0.164	0.147	0.038	1	0.05	1
6BA	0.211	0.189	0.050	1	0.05	1
5BA	0.239	0.215	0.056	1	1/16	1
4BA	0.269	0.243	0.063	1	1/16	1
3BA	0.307	0.277	0.072	1	5/64	1
2BA	0.351	0.319	0.083	1	3/32	1

Notes: socket-head cap and countersunk screw tables

1. Length (L) is the maximum length screw made for the given thread size.

2. Screws of a length equal to, or shorter than, the dimension given in column L, are threaded up to the head. Dimension T only applies to screws longer than the length given in column L.

3. For screws longer than the dimension given in column L, dimension T applies but the value given is the minimum value and varies by up to +¼in, depending on the length of screw.

4. Maximum diameter at theoretical sharp corner.

HEXAGON-HEAD NUTS AND SCREWS

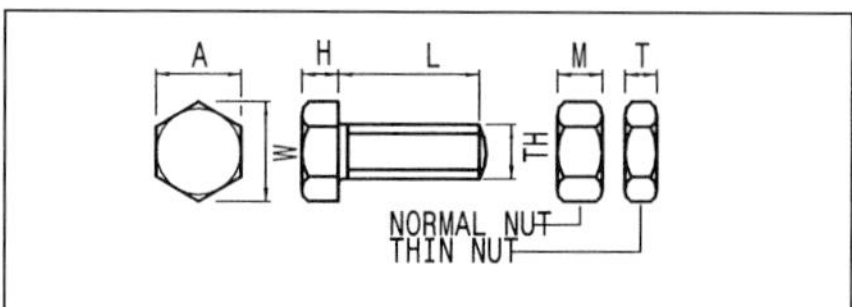

Table 21.8. ISO metric hexagon-head nuts and screws: sizes and dimensions (mm)

TH	A	H	W	M	T
M1.6	3.2	1.22	3.7	1.3	-
M2	4.0	1.52	4.6	1.6	-
M2.5	5.0	1.82	5.8	2.0	-
M3	5.5	2.12	6.4	2.4	-
M4	7.0	2.92	8.1	3.2	-
M5	8.0	3.65	9.2	4.0	-
M6	10.0	4.15	11.5	5.0	-
M8	13.0	5.65	15.0	6.5	5
M10	17.0	7.18	19.6	8.0	6
M12	19.0	8.18	21.9	10.0	7
M16	24.0	10.18	27.7	13.0	8
M20	30.0	13.21	34.6	16.0	9

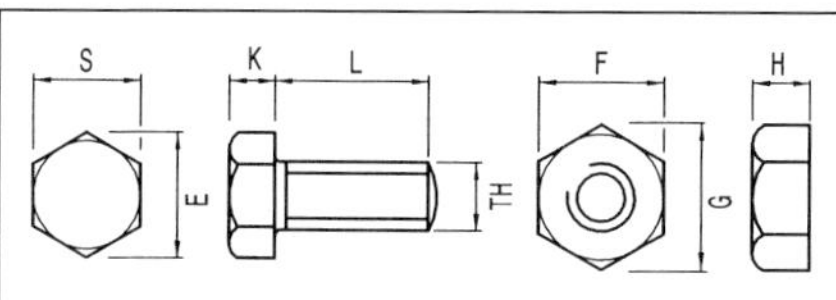

Table 21.9. UNF/UNC (machine screws): sizes and dimensions (in)

TH	S(2)	K	E	F(2)	H	G
1	0.125	0.044	0.134	0.156	0.050	0.180
2	0.125	0.050	0.134	0.188	0.066	0.217
3	0.188	0.055	0.202	0.188	0.066	0.217
4	0.188	0.060	0.202	0.250	0.098	0.289
5	0.188	0.070	0.202	0.312	0.114	0.361
6	0.250	0.093	0.272	0.312	0.114	0.361
8	0.250	0.110	0.272	0.344	0.130	0.397
10	0.312	0.120	0.340	0.375	0.130	0.433
12	0.312	0.155	0.340	0.438	0.161	0.505

TH	S(2)	K	E	F(2)	H	G
1/4	0.375	0.190	0.409	0.438	0.193	0.505
5/16	0.500	0.230	0.545	0.562	0.225	0.650
3/8	0.562	0.295	0.614	0.625	0.257	0.722

Table 21.0. UNF/UNC (standard hexagon screws): sizes and dimensions (in)

TH	S(2)	K	E	F(2)	H	G
1/4	0.438	0.163	0.505	0.438	0.226	0.505
5/16	0.500	0.211	0.577	0.500	0.273	0.577
3/8	0.562	0.243	0.650	0.562	0.337	0.650
7/16	0.625	0.291	0.722	0.688	0.385	0.794
1/2	0.750	0.323	0.866	0.750	0.448	0.866
9/16	0.812	0.371	0.938	0.875	0.496	1.010

Screw lengths, metric

The preferred lengths for metric screws are 2.5, 3, 4, 5, 6, 8, 10, 12, 14, 16 and 20mm, then at 5mm increments to 90mm.

The recommended range for each diameter is:

M2	2.5–20mm
M2.5	3–25mm
M3	4–30mm
M4	5–40mm
M5	6–50mm
M6	8–60mm
M8	10–80mm
M10	12–100mm

Notes: hexagon-head screws and nuts tables

1. For UNF/UNC hexagon-head screws and nuts there are alternative across-flats dimensions for some thread sizes, known as large-duty or heavy-duty. These alternatives are not listed.

2. The across-flats dimensions for UNF/UNC nuts are in some cases larger than those for the head of a screw with the same thread size, mostly for the

numbered sizes. Screw and nut across-flats dimensions are given separately.

3. All screws and nuts are made to a tolerance and the values quoted are the nominal sizes. In most cases, these are the maximum value.

CIRCLIPS

Circlips are made in three main forms: external, internal and E clip series (see drawings and tables below). Small variations in shape are permitted, particularly at the smaller and larger sizes.

EXTERNAL SERIES

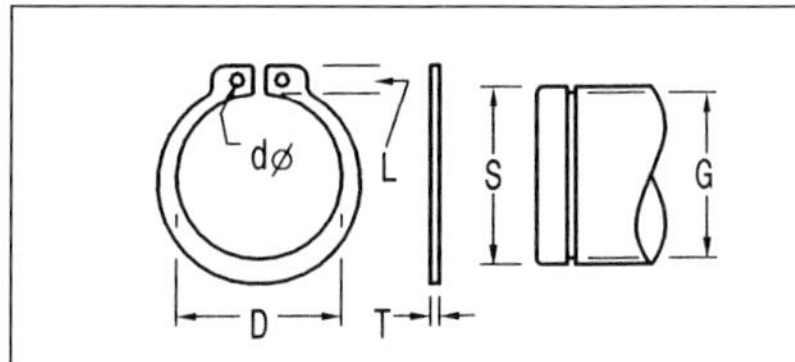

Table 21.11. External circlip series: sizes and dimensions (mm)

R	Shaft		Clip			
	S	G	D	T	d	L
S003M	3	2.8	2.66	0.4	0.8	1.9
S004M	4	3.8	3.64	0.4	1.0	2.2
S005M	5	4.8	4.64	0.6	1.0	2.5
S006M	6	5.7	5.54	0.7	1.15	2.7
S007M	7	6.7	6.45	0.8	1.2	3.1
S008M	8	7.6	7.35	0.8	1.2	3.2
S009M	9	8.6	8.35	1.0	1.2	3.3
S010M	10	9.6	9.25	1.0	1.5	3.3
S011M	11	10.5	10.2	1.0	1.5	3.3
S012M	12	11.5	11.0	1.0	1.7	3.3
S013M	13	12.4	11.9	1.0	1.7	3.4
S014M	14	13.4	12.9	1.0	1.7	3.5
S015M	15	14.3	13.8	1.0	1.7	3.6

R	Shaft		Clip			
	S	G	D	T	d	L
S016M	16	15.2	14.7	1.0	1.7	3.7
S017M	17	16.2	15.7	1.0	1.7	3.8
S018M	18	17.0	16.5	1.2	2.0	3.9
S019M	19	18.0	17.5	1.2	2.0	3.9
S020M	20	19.0	18.5	1.2	2.0	4.0
S021M	21	20.0	19.5	1.2	2.0	4.1
S022M	22	21.0	20.5	1.2	2.0	4.2
S023M	23	22.0	21.5	1.2	2.0	4.3
S024M	24	22.9	22.2	1.2	2.0	4.4
S025M	25	23.9	23.2	1.2	2.0	4.4
S026M	26	24.9	24.2	1.2	2.0	4.5
S027M	27	25.6	24.9	1.2	2.0	4.6
S028M	28	26.6	25.9	1.5	2.0	4.7
S029M	29	27.6	26.9	1.5	2.0	4.8
S030M	30	28.6	27.9	1.5	2.0	5.0
S031M	31	29.3	28.6	1.5	2.5	5.1
S032M	32	30.3	29.6	1.5	2.5	5.2
S033M	33	31.3	30.5	1.5	2.5	5.3
S034M	34	32.3	31.5	1.5	2.5	5.4
S035M	35	33.0	32.2	1.5	2.5	5.6
S036M	36	34.0	33.2	1.75	2.5	5.6
S037M	37	35.0	34.2	1.75	2.5	5.7
S038M	38	36.0	35.2	1.75	2.5	5.8

Notes

1. The recommended width of the groove is 0.1mm greater than the width of the circlip.

2. Diameter G is larger than the free internal diameter of the clip, thus ensuring that the clip is held firm when fitted. It should be within +0–0.1mm up to S018M and +0–0.2mm above S018M to guarantee this.

3. Key to abbreviations:

R	circlip reference (external series)
S	shaft diameter
G	groove diameter
D	inside diameter
T	thickness
d	lug hole diameter
L	lug depth

INTERNAL SERIES

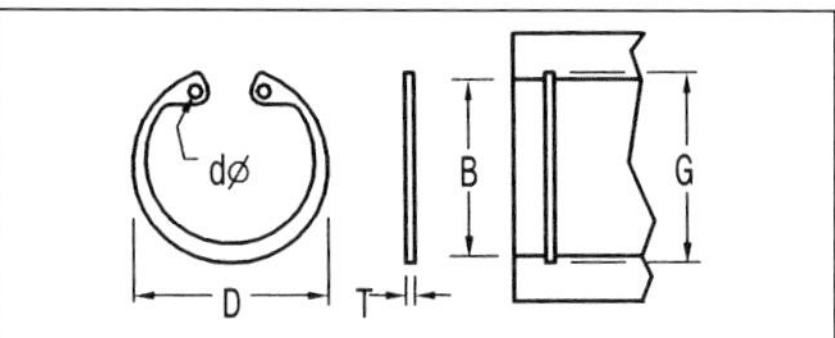

Table 21.12. Circlips, internal series: sizes and dimensions (mm)

	Bore		Clip			
R	B	G	D	T	d	L
B008M	8	8.4	8.7	0.8	1.0	2.4
B009M	9	9.4	9.8	0.8	1.0	2.5
B010M	10	10.4	10.8	1.0	1.2	3.2
B011M	11	11.4	11.8	1.0	1.2	3.3
B012M	12	12.5	13.0	1.0	1.5	3.4
B013M	13	13.6	14.1	1.0	1.5	3.6
B014M	14	14.6	15.1	1.0	1.7	3.7
B015M	15	15.7	16.2	1.0	1.7	3.7
B016M	16	16.8	17.3	1.0	1.7	3.8
B017M	17	17.8	18.3	1.0	1.7	3.9
B018M	18	19.0	19.5	1.0	2.0	4.1
B019M	19	20.0	20.5	1.0	2.0	4.1
B020M	20	21.0	21.5	1.0	2.0	4.2
B021M	21	22.0	22.5	1.0	2.0	4.2
B022M	22	23.0	23.5	1.0	2.0	4.2

Notes

1. The recommended width of the groove is 0.1mm greater than the width of the circlip.

2. Diameter G is smaller than the free external diameter of the clip, thus ensuring that the clip is held firm when fitted. It should be within +0.1−0mm up to B018M and +0.2−0mm above B018M to guarantee this.

3. Key to abbreviations:

R	circlip reference (internal series)
B	bore diameter
G	groove diameter
D	circlip diameter
T	thickness
d	lug hole diameter
L	lug depth

E CLIP SERIES

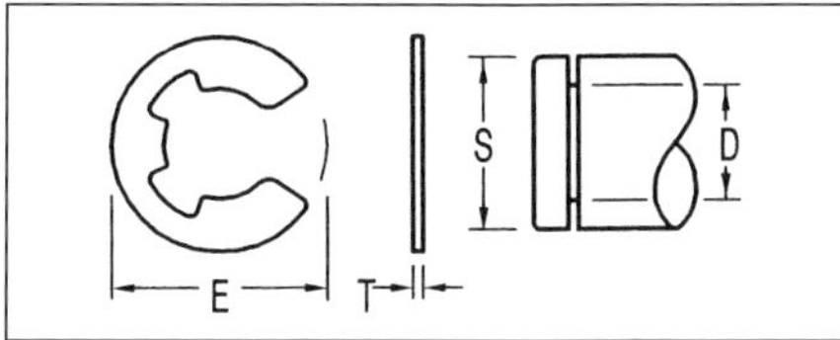

Table 21.13. Circlips, E clip series: sizes and dimensions (mm)

	Shaft			Clip	
R	S (min.)	S (max.)	D	E	T
008MS	1.0	1.4	0.8	2.0	0.22
012MS	1.4	2.0	1.2	3.0	0.32
015MS	2.0	2.5	1.5	4.0	0.42
019MS	2.5	3.0	1.9	4.5	0.52
023MS	3.0	4.0	2.3	6.0	0.62
032MS	4.0	5.0	3.2	7.0	0.62
040MS	5.0	7.0	4.0	9.0	0.72
050MS	6.0	8.0	5.0	11.0	0.72
060MS	7.0	9.0	6.0	12.0	0.72
070MS	8.0	11.0	7.0	14.0	0.92

	Shaft			Clip	
R	S (min.)	S (max.)	D	E	T
080MS	9.0	12.0	8.0	16.0	1.03
090MS	10.0	14.0	9.0	18.5	1.13
100MS	11.0	15.0	10.0	20.0	1.23
120MS	13.0	18.0	12.0	23.0	1.33
150MS	16.0	24.0	15.0	29.0	1.53
190MS	20.0	25.0	19.0	37.0	1.78
240MS	25.0	38.0	24.0	44.0	2.03

TAPER PINS

These are used primarily for fixing items such as gears, pulleys and brushes to sharfts as indicated in the following sketch.

TAPER

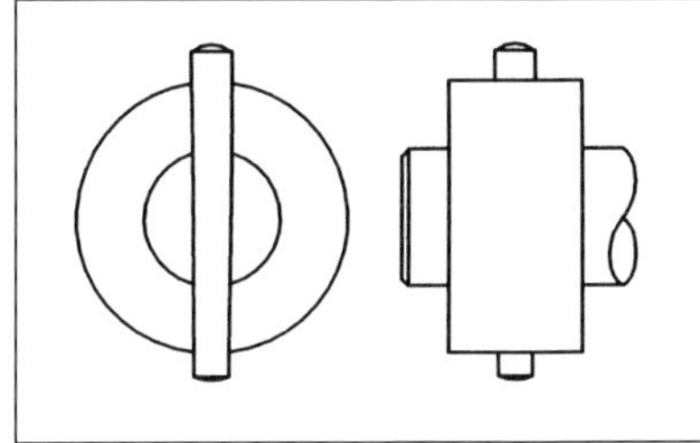

The taper for imperial pins is 1 in 48 on diameter, while that for metric pins is 1 in 50 on diameter. As a result different reamers are required for each system.

SIZES

The sizes quoted for imperial and metric taper pins are based on different dimensions: imperial pins are known by the nominal diameter at the larger end whereas metric pins are known by the nominal diameter at the smaller end. This is a common source of error, so orders for pins and/or reamers should be very specific. Imperial pins and reamers were once known by a reference number (see 'No.' in Table 21.14); this is now non-preferred.

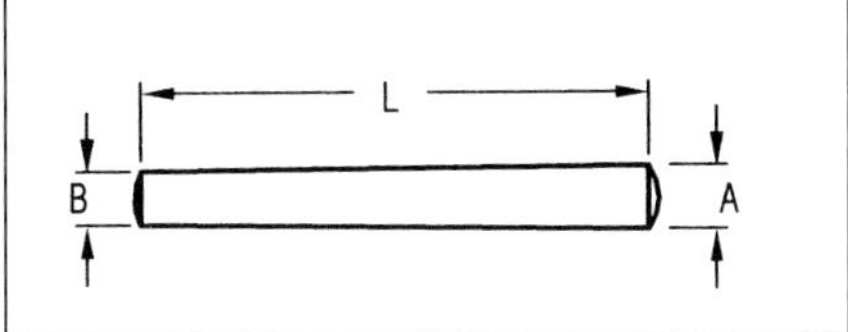

LENGTHS

Tables 21.14 and 21.15 give a representative range of lengths. The full metric range is 4mm, 5mm, 6mm to 32mm, in steps of 2mm, followed by 35mm and up to 55mm in steps of 5mm, the lengths available depending on the diameter. Diameters larger than 4mm are also made in longer lengths.

REFERENCE CODE FOR METRIC PINS

The reference code for a metric taper pin is given as, e.g., A 6 × 30 ST, where A indicates a ground pin (or B if it is a turned pin), 6 is the nominal diameter at the small end, 30 is the length and ST indicates that the pin is made from unhardened free-cutting steel.

TAPER-PIN REAMERS

As for taper pins (see above), the sizes of metric and imperial taper-pin reamers are based on different dimensions. Metric reamers are defined by their nominal diameter at the smaller end, whereas imperial reamers are defined by the nominal diameter at the larger end. Both are available but care should be taken when ordering them. Imperial reamer sizes were once known by a number (see 'No' in Table 21.16), which is now non-preferred. Metric and imperial reamers are

Table 21.14. Imperial taper pins: dimensions (in)

Nom. dia.	Dia. A	Diameter B for typical lengths (L)					
		0.50	0.75	1.00	1.25	1.50	No.
1/16	0.0625	0.052	0.047	0.042	–	–	7/0
5/64	0.0780	0.068	0.062	0.057	–	–	6/0
3/32	0.0940	0.084	0.078	0.073	–	–	5/0
7/64	0.1090	0.099	0.093	0.088	–	–	4/0
1/8	0.1250	0.115	0.109	0.104	–	–	3/0
9/64	0.1410	0.131	0.125	0.120	0.115	–	2/0
5/32	0.1560	0.146	0.140	0.135	0.130	–	0
11/64	0.1720	–	0.156	0.151	0.146	–	1
3/16	0.1930	–	0.177	0.172	0.167	0.162	2

Table 21.15. Metric taper pins: dimensions (mm)

Nom. dia.	Dia. B	Diameter A for typical lengths (L)					
		6	10	16	22	30	40
0.6	0.6	0.72	0.80	–	–	–	–
0.8	0.8	0.92	1.00	–	–	–	–
1.0	1.0	1.12	1.20	1.32	–	–	–
1.2	1.2	1.32	1.40	1.52	1.64	–	–
1.5	1.5	1.62	1.70	1.82	1.94	–	–
2	2	–	2.20	2.32	2.44	2.60	–
2.5	2.5	–	2.70	2.82	2.94	3.10	3.30
3	3	–	3.20	3.32	3.44	3.60	3.80
4	4	–	–	4.32	4.44	4.60	4.80

not interchangeable because their tapers differ, metric being 1 in 50 on diameter and imperial being 1 in 48.

SIZES

Sizes are designed to correspond to pin sizes. However, there is some overlap between the diameters of adjacent sizes of reamer, so a reamer of a given size may be adequate for a short pin of an adjacent size.

Table 21.16. Imperial reamers: sizes and dimensions (in)

Size	Large dia.	Small dia.	No.
1/16	0.064	0.043	7/0
5/64	0.080	0.059	6/0
3/32	0.095	0.069	5/0
7/64	0.111	0.080	4/0
1/8	0.127	0.090	3/0
9/64	0.143	0.104	2/0
5/32	0.158	0.116	0
11/64	0.174	0.127	1
3/16	0.195	0.138	2

Table 21.17 Metric reamers: sizes and dimensions (mm)

Size	Large dia.	Small dia.
0.6	0.90	0.5
0.8	1.18	0.7
1.0	1.46	0.9
1.2	1.74	1.1
1.5	2.14	1.4
2.0	2.86	1.9
2.5	3.36	2.4
3.0	4.06	2.9

O RINGS

SIZES

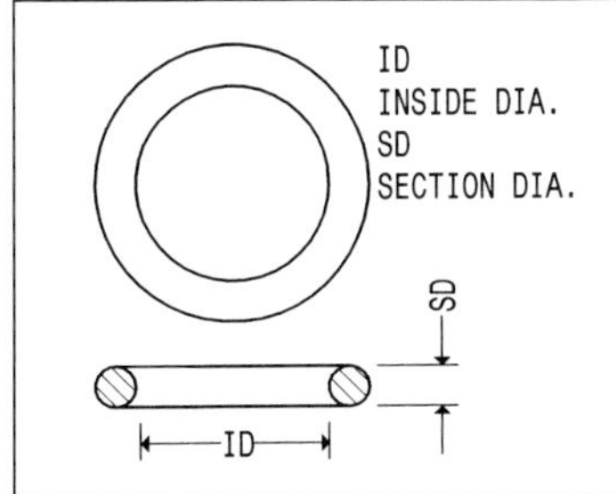

As the drawing above shows, only two dimensions are used to define the O ring: inside diameter (ID) and section diameter (SD).

Metric rings are available in SDs of 1.6, 2.4, 3.0, 5.7 and 8.5mm. IDs range from 3 mm to 499mm, with increments of 1–20mm, depending on the SDs and IDs. For example, at an SD of 1.6mm, increments are about 1–3mm, while at an SD of 5.7mm, increments are about 2–20mm, with the larger increments at the large IDs.

Table 21.18. O rings: section and inside diameters (mm)

SD	ID range
1.6	3.1–37.1
2.4	3.6–69.6
3.0	19.5–249.5
5.7	44.3–499.3
8.4	144.1–249.1

MATERIALS

Rings are made from a wide range of materials, each having characteristics that make it suitable for very specific applications. Typically, the materials used are: ethylene-propylene, natural rubber, nitrile-acrylonitrile-butadiene, polyurethane and silicone. This list is far from exhaustive and, in any case, there can be different grades of a particular material.

The different materials are designed to cope with requirements such as: high temperature, low temperature, resistance to particular chemicals and ability to work in dynamic applications, i.e. rotating or reciprocating spindles.

Polyurethane rings are often used as drive belts in transmission systems.

DYNAMIC APPLICATIONS

Not all sizes of rings are considered suitable for reciprocating applications, and only a limited range of the smaller IDs are recommended for such duties. Table 21.19 lists the sizes that are considered acceptable, although some manufactures give a slightly wider range, with a few larger diameters in the case of 3.0 and 5.7 rings. The rings listed can also be used for static applications.

Table 21.19. O rings:
acceptable sizes for dynamic applications (in)

SD	ID range
1.6	None
2.4	3.6–18.6
3.0	19.5–42.5
5.7	44.3–119.3
8.4	144.1– 249.1

All ring sizes are regarded as acceptable for low-speed rotating applications, with peripheral speed being the limiting factor. Therefore rings with a smaller ID can run at higher speeds than those with a larger ID.

The material from which the ring is made, together with the surface finish of the mating parts, will have a bearing on what is permissible.

DESIGN CONSIDERATIONS

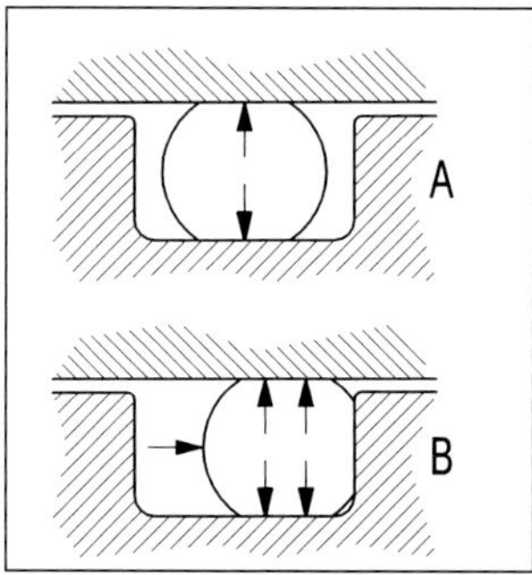

In the above drawing, example A shows an O ring in its assembled but unloaded state, where the width of the groove in which it sits is wider than the SD of the ring.

Example B shows that, when loaded by the pressure of the mediums, the ring will be forced against one wall of the groove, so that it attempts to expand, thus increasing the pressure (as indicated by the double arrows) and further improving the seal. This is assuming that the initial compression at A

is sufficient to prevent the working pressure by-passing the ring.

Groove features

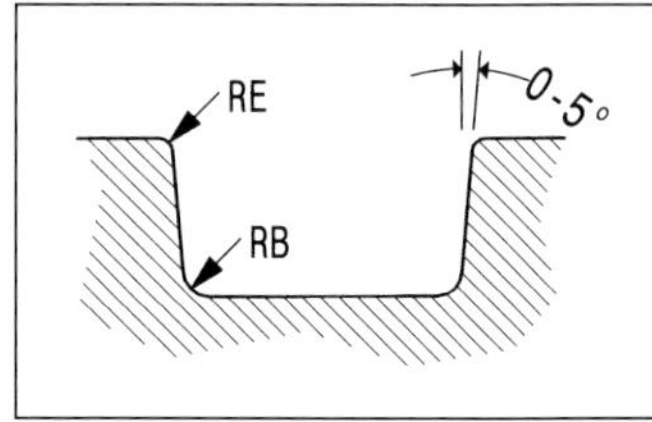

Rather than the groove in which the ring sits being a simple rectangle other features are included. The most important is radius RE, which ensures that the ring is not damaged as it drops into the groove, which could easily happen if the edge was left sharp.

Radius RB is perhaps a little less important but avoids the need for the cutting tool to have a sharp corner that may chip. In very demanding applications a sharp corner could result in a fracture point, and the radius considerably reduces this possibility.

The slope to the side of the groove, of up to 5°, is no doubt to assist in the part's production.

Table 21.20. Groove radii: dimensions (in)

SD	RE (max.)	RB (max.)
1.6	0.2	0.20/0.40
2.4	0.5	0.20/0.40
3.0	1.0	0.20/0.40
5.7	1.0	0.20/0.40
8.4	1.0	0.20/0.40

For simplicity the radii and the sloping side have been omitted from subsequent drawings.

'O' ring assemblies

For clarity, the following drawings show the assembly without the ring being fitted.

Ring in spindle

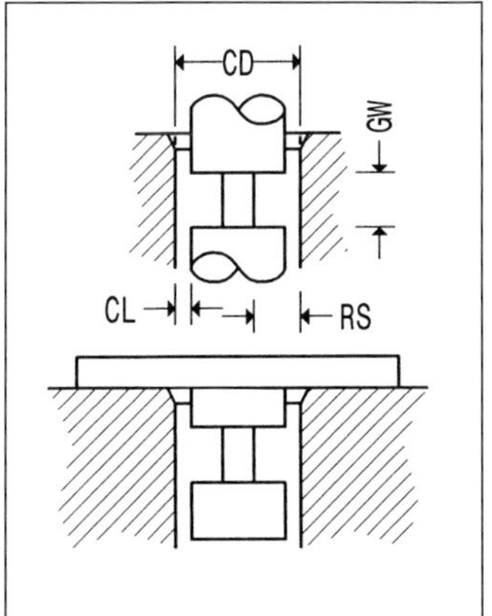

The drawing above shows the most common arrangement for the use of an O ring. Note that the ring space (RS) is not the depth of the groove in the spindle but includes the clearance (CL). The depth of the groove in the spindle will vary according to the chosen clearance.

Table 21.21. Ring in spindle: dimensions (in)

Section	Groove	Depth	Clearance
SD	GW	RS	CL (max.)
1.6	2.3/2.5	1.18/1.25	0.06
2.4	3.2/3.4	1.97/2.09	0.07
3.0	4.0/4.2	2.50/2.65	0.08
5.7	7.5/7.7	4.95/5.18	0.09
8.4	11.0/11.2	7.50/7.75	0.10

Ring in cylinder

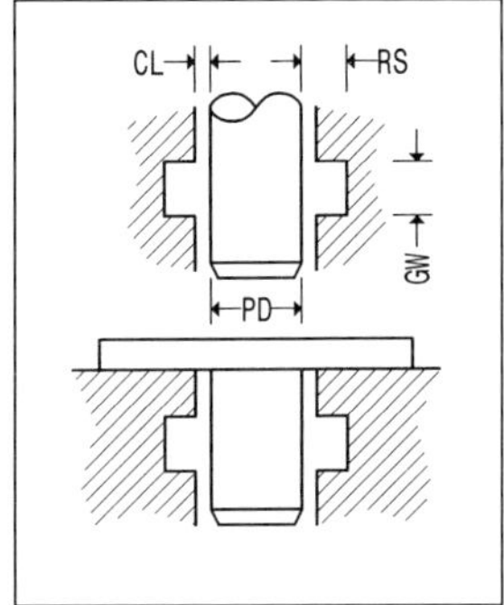

See 'Ring in Spindle' for comments and dimensions.

Ring in triangular groove

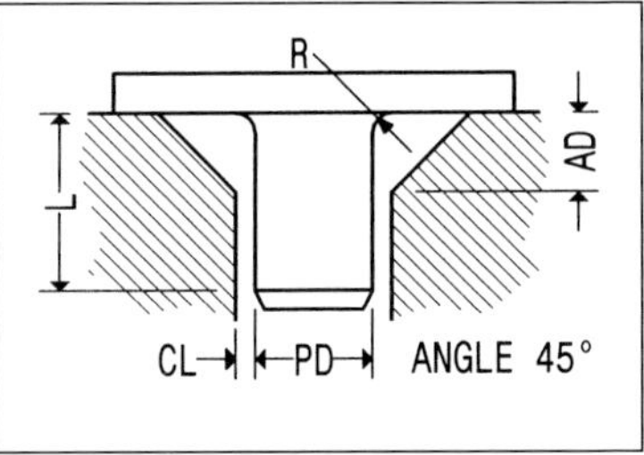

Table 21.22. Ring in triangular groove: dimensions (in)

SD	CL	AD	R (max.)	L (min.)
1.6	0.06	2.20/2.32	0.8	4
2.4	0.07	3.30/3.42	1.3	5
3.0	0.08	4.20/4.32	2.0	6
5.7	0.09	7.80/7.92	3.0	10
8.4	0.10	11.50/11.62	4.0	14

It is difficult to make the chamfer accurate enough to ensure a reliable seal so care should be taken when adopting this method.

Ring in flange

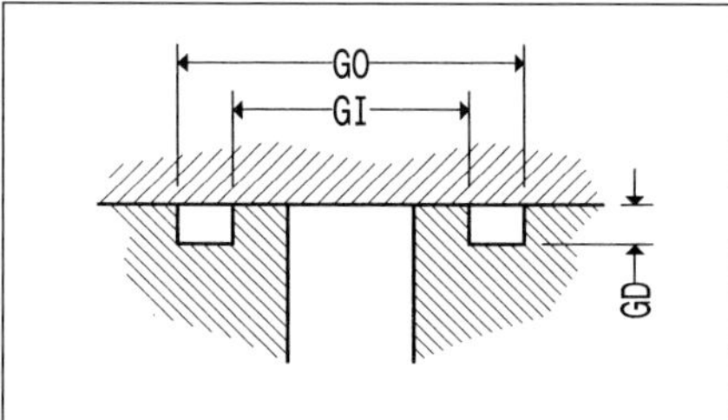

This arrangement is more complex than those mentioned as the dimensions depend on whether the higher pressure is inside or outside of the cylinder. If it is inside, the ring will be forced against the outside diameter of the groove whereas, if it is outside, the reverse will be the case.

Table 21.23. Groove depth (in)

SD	GD
1.6	1.2/1.3
2.4	1.7/1.8
3.0	2.2/2.3
5.7	4.4/4.5
8.4	6.6/6.7

ID dependence

The details given so far are independent of the ID of the ring being used, but items such as cylinder diameter, shaft diameter, etc. must relate to the size of the associated 'O' ring. With so many different sizes of ring, it is quite impossible to list them all but, fortunately, the relationship between the ring and its mating part is a constant, except at very small and very large diameter rings. Even here the variations are very small and, therefore, except for critical applications, the following approach should suffice.

Table 21.24 gives a single ID size for each O-ring SD, together with its required dimensions. To arrive at the dimensions for other diameters, all dimensions should be adjusted by the same margin, e.g. if using a ring with a diameter 20mm larger than that listed, increase all dimensions by 20mm, while if using a ring with a diameter 5mm smaller, then reduce all dimensions by 5mm.

Table 21.24. Flange groove diameters (in)

		Pressure			
		Internal		External	
SD	ID	GI (max.)	GO	GI	GO (min.)
1.6	22.1	21.0	25.3	22.5	26.5
2.4	20.6	18.5	25.4	21	27
3.0	21.5	19	27	22	30
5.7	44.3	41	55	45	59
8.4	144.1	140	160	145	165

Table 21.25. Shaft/Cylinder diameters (in)

SD	ID	PD	CD
1.6	22.1	22.5	25
2.4	20.6	21	25
3.0	21.5	22	27
5.7	44.3	45	55
8.4	144.1	145	160

Assembly

Some of the drawings show a chamfered end to the shaft the purpose of which is to aid assembly and prevent the ring becoming damaged as it may be if it were forced onto a spindle having a sharp edge to its end. Similarly, where the ring is first assembled into a groove on the side of a cylinder this also has a chamfered entry for the same reasons.

The angle of the chamfer should be in the region of 15–20° and, if on the end of a shaft, the diameter of the smallest end should be a little smaller than the inside diameter of the ring being fitted. Likewise, when being fitted into a cylinder, the diameter of the larger end should be a little larger than the outer diameter of the ring.

If the ID is less than three times the SD, any attempt to fit the ring into a conventional groove may damage the ring, due to the difficulty of expanding the ring sufficiently. In this case the component would need to be made in two parts, by splitting it at the ring's housing. This is not easy and the situation is best avoided.

Imperial size rings

These are made in SD sizes of 0.070, 0.103, 0.139, 0.210 and 0.275in. There is a very much larger number of ID sizes than with metric rings.

The above should give the reader an understanding of the important factors when designing an O-ring assembly. However, for critical applications, advice should be sought from the ring manufacturers.

SELF-TAPPING SCREWS

Just as slotted and hexagon-headed screws have been added too in recent years by drive methods such as Phillips, Pozi, Torx, etc., manufacturers have developed many forms of self-tapping screw. However, like the drive methods, only a few self-tapping screw types are commonly used.

FORMS AND SIZES

The drawing shows types AB and B which are the most common forms. Type A, also shown, has been made in the the same size as type AB but with a coarser thread, it has though, been largely superseded by type AB.

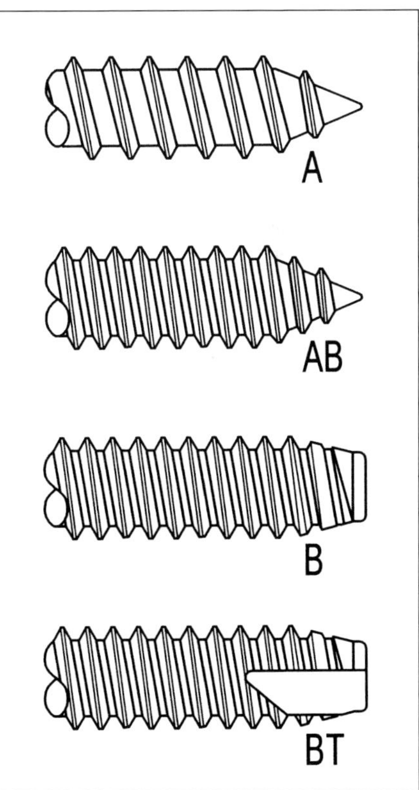

IMPERIAL SCREWS

Table 21.26. Imperial screws: sizes and dimensions (in)

No.	OD	TPI
2	0.086	32
4	0.112	24
6	0.138	20
8	0.164	18
10	0.190	16
12	0.214	14
14	0.250	14

Table 21.27. Metric screws: sizes

Dia.		Pitch
2.2	×	0.8
2.9	×	1.0
3.5	×	1.3
4.2	×	1.4
4.8	×	1.6
5.5	×	1.8
6.3	×	1.8

As Tables 21.26 and 21.27 show, imperial sizes are referred to by a number while metric sizes are referred to by diameter and pitch. However, if the metric diameter and pitch are converted to imperial diameter and TPI, it will be seen that both ranges are virtually identical. These very small differences may be due to allowances for manufacturing tolerances.

HOLE SIZES

The hole sizes listed in Table 21.28 are for use when mounting screws into steel. However, as there are different grades of steel, they should only be used for guidance and, wherever possible, should be checked with a test piece before they are applied to the assembly being made.

Table 21.28. Hole sizes for mounting into steel

Screw size		Thickness	Hole dia.	Alu.
Imperial	Metric	mm	mm	Y/N
2	2.2	0.45	1.60	N
–	–	0.71	1.70	Y
–	–	0.91	1.85	Y
–	–	1.22	1.90	Y
–	–	1.62	1.95	Y
4	2.9	0.45	2.05	N
–	–	0.71	2.20	N
–	–	0.91	2.30	Y
–	–	1.22	2.35	Y
–	–	1.62	2.40	Y
–	–	2.03	2.60	Y
6	3.5	0.45	2.35	N
–	–	0.71	2.60	N
–	–	0.91	2.80	Y
–	–	1.22	2.85	Y
–	–	1.62	2.95	Y
–	–	2.03	3.10	Y
–	–	2.64	3.20	Y
8	4.2	0.71	2.90	N
–	–	0.91	3.10	Y
–	–	1.22	3.20	Y
–	–	1.62	3.40	Y
–	–	2.64	3.70	Y
–	–	3.18	3.80	Y
10	4.8	0.71	3.40	N
–	–	0.91	3.20	Y
–	–	1.22	3.60	Y
–	–	1.62	3.80	Y
–	–	2.64	4.10	Y
–	–	3.18	4.30	Y
–	–	4.75	4.50	Y
12	5.5	0.71	4.10	N
–	–	0.91	4.20	N
–	–	1.22	4.30	Y
–	–	1.62	4.50	Y
–	–	2.64	4.80	Y
–	–	3.18	4.90	Y

Screw size		Thickness	Hole dia.	Alu.
Imperial	Metric	mm	mm	Y/N
–	–	4.75	5.10	Y
14	6.3	1.22	4.80	N
–	–	1.62	5.20	Y
–	–	2.64	5.50	Y
–	–	3.18	5.70	Y
–	–	4.75	5.90	Y
–	–	6.35	6.00	Y

The last column (Alu.) of Table 21.28 indicates whether or not it is acceptable (Yes or No) to use the screw in aluminum of the given thickness. If strength is important the hole size should be reduced, say by 0.1mm at the smaller screw sizes and up to 0.5mm smaller at the largest.

BT TYPE SCREWS

These screws are dimensionally identical to type B but with a short cutting edge, thereby producing a predominantly cut thread. They are not intended for use in steel components or thinner sheet materials.

Their main use is in metals such as die-cast aluminum and zinc, but also in plastics. This wide diversity of material types means that the use of a test piece to determine hole size is highly recommended, especially in critical applications.

Hole sizes

The cutting action of these screws results in only a small variation in hole size for differing depths of thread (in the order of 0.1mm) so only one hole size is given for each screw diameter. Also, because of their cutting action, they can be used to a greater depth than the thread-forming B-type screws. Very approximately, the range is from 60% to 200% of the screw diameter for metal components and, in the case of plastic components, from 150% to 300%.

Table 21.29. Hole sizes for cast zinc/aluminum

Screw size		Hole dia.
Imperial	Metric	mm
2	2.2	1.9
4	2.9	2.5
6	3.5	3.1
8	4.2	3.8
10	4.8	4.3
12	5.5	4.9
14	6.3	5.8

Table 21.30. Hole sizes for plastics, acrylic, etc.

Screw size		Hole dia.
Imperial	Metric	mm
2	2.2	1.9
4	2.9	2.5
6	3.5	3.1
8	4.2	3.7
10	4.8	4.2
12	5.5	4.8
14	6.3	5.6

STANDARD THREAD FORM SELF-TAPPING SCREWS

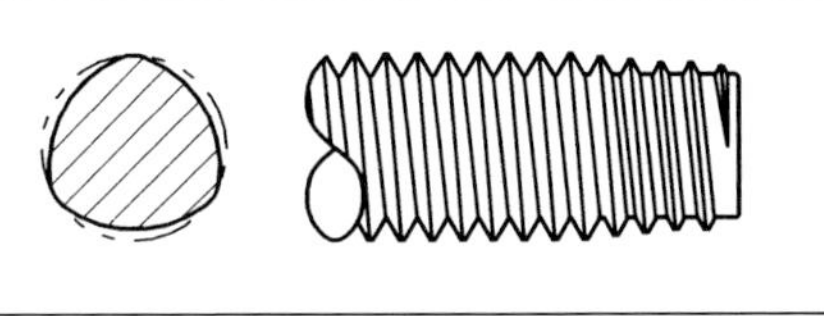

The trilobular screw shown in the drawing is a quite different form of self-tapping screw that produces a standard thread form, i.e. Metric, Unified, etc. The advantages of this type of screw are:

- stronger assembly
- being virtually the equal of a standard screw
- easy dismantling and re-assembly
- a standard screw can be used in its place if it is lost in the process.

Also, when used in relatively thin metal, the formed thread has a degree of work hardening and is therefore stronger than a cut thread. There are also no partial threads as might be likely if a tap failed to start the hole cleanly.

Although it is not their intended purpose, a single screw will produce many tapped holes and, if required in a thin sheet, will produce a better result than a thread cutting tap. Standard screws can then be used in the final assembly. The screws do not look that different from a standard screw but, if rolled between the thumb and first finger, the trilobular form will be apparent, even on the smaller sizes.

Hole sizes

Table 21.31 gives a suggested hole size for each size screw and the percentage of a full thread achieved. As with conventional tapping, the percentage depth will vary considerably when moving up or down a standard drill size, especially at the smaller sizes. This is quantified by the values given with changes of ±0.1mm in the drill size. However, drills are available in 0.05mm increments at the smaller sizes and a drill of 2.75mm will give a depth of about 75% in the case of the M3 size.

In mass production, holes will often be punched so hole sizes will not be governed by standard drill sizes. Table 21.31 therefore also includes the hole size required for a 70% thread depth.

Table 21.31. Hole sizes and percentage of full thread per size of screw

Thread	Hole	%	% at −0.1mm	% at +0.1mm	Hole at 70%
M3 × 0.5	2.8	60	91	35	2.77
M4 × 0.7	3.7	65	87	40	3.68
M5 × 0.8	4.6	76	96	55	4.64
M6 × 1.0	5.5	76	92	60	5.54
M8 × 1.25	7.4	74	86	60	7.43
M10 × 1.5	9.3	72	82	60	9.32

As indicated in Chapter 8, a thread depth of between 60% and 70% will be strong enough for most applications and reduces the possibility of broken taps. This approach will be equally acceptable with these screws and, in this case, will reduce the torque required to drive the screw home. As the thread is produced by metal extrusion, the actual thread depth will be rather more than the theoretical values given.

The drawing shows that the first few threads (between three and five depending on size) are smaller in diameter, the purpose of which is to assist in getting the screw started. When used in sheet material these smaller threads should pass completely through the sheet. This is, of course, not possible with blind holes, where a length of full thread at least equal to the screw's diameter should be engaged with the component.

As with other self-tapping screws, the suitability of the proposed hole size should be checked in a test piece before committing it to the item being made.

CHAPTER 22
MATERIAL DIMENSIONS

WIRE GAUGES

SYSTEMS

There have been so many systems for wire diameters that it is impossible to provide an all-embracing list. The following tables include what may be considered as the main systems.

However, when ordering materials, it is best to quote diameter in terms of inches or metric units to avoid any misunderstanding. For example, where an early document quotes a diameter by system and number (e.g. as 16SWG) the table below (Table 22.1) gives the diameter in inches and millimeters. Be aware that names (and initials, e.g. BWG) of the systems have not always been used consistently.

System names

Unfortunately, some ranges have been known by more than one name. The following list should clarify the common ones:

- SWG (Standard wire gauge): also known as Imperial wire gauge.
- W&M (Washburn & Meon): also known as SWG (Steel Wire Gauge; US).
- BWG (Birmingham wire gauge): also known as Stubs Iron Wire. This should not be confused with BG (Birmingham gauge) which is for sheet materials.
- AWG (American Wire Gauge): also known as Brown & Sharpe Wire Gauge.
- EMW and ASW: see Piano wire below.

Metrication

Metric size wires are intended to be based on the ISO preferred numbers series, as explained in Chapter 24.

Therefore the table also includes the R40 number series for comparison with earlier wire sizes, around which a standard is likely to be finalized. The lower number series, R10 and R20, will be the basis for the preferred values but no doubt manufacturers who are continuing to make to old designs will attempt to use the nearest value to that used in the old wire size ranges. This will inevitably produce some values from the R40 range of numbers, and only time will tell what sizes will eventually become preferred.

PIANO WIRE

Piano wire (also called music wire in some listings) has been made to a wide range of standard diameters, probably more so than plain steel and copper wires. This makes it impossible to give piano wire sizes with any certainty. Because of this only two systems are listed, i.e.:

- EMW: English music wire gauge
- ASW: American Steel and Wire Co. music wire gauge.

Even with these, the descriptions found on old designs are more likely to refer to, e.g., 10G Piano Wire. In this case, you can only assume that this is EMW if it is a UK design or ASW if it is a US design.

Table 22.1. Wire gauges: comparison

SWG	W&M	BWG	AWG	EMW (UK)	ASW (US)	Metric R40	in	mm
-	47	35	36	-	-	-	0.0050	0.127
39	46	-	-	-	-	0.132	0.0052	0.132
-	45	-	-	-	-	0.140	0.0055	0.140
-	-	-	35	-	-	-	0.0056	0.142
-	44	-	-	-	-	-	0.0058	0.147
-	-	-	-	-	-	0.150	0.0059	0.150
38	43	-	-	-	-	-	0.0060	0.152
-	42	-	-	-	-	-	0.0062	0.157
-	-	-	34	-	-	0.160	0.0063-	0.160
-	41	-	-	-	-	-	0.0066	0.168
-	-	-	-	-	-	0.170	0.0067	0.170
37	-	-	-	-	-	-	0.0068	0.173
-	40	34	-	-	-	-	0.0070	0.178
-	-	-	33	-	-	0.180	0.0071	0.180
-	39	-	-	-		0.190	0.0075	0.190
36	-	-	-	-		-	0.0076	0.193
-	-	-	-	-	-	0.200	0.0079	0.200
-	38	33	32	-	-	-	0.0080	0.203
-	-	-	-	-	-	0.212	0.0083	0.212
35	-	-	-	-	-	-	0.0084	0.213
-	37	-	-	-	-	-	0.0085	0.216
-	-	-	-	-	-	0.224	0.0088	0.224
-	-	-	31	-	-	-	0.0089	0.226
-	36	32	-	-	-	-	0.0090	0.229
34	-	-	-	-	-	-	0.0092	0.234
-	-	-	-	-	-	0.236	0.0093	0.236
-	35	-	-	-	-	-	0.0095	0.241
-	-	-	-	-	-	0.250	0.0098	0.250
33	-	31	30	-	1	-	0.0100	0.254
-	34	-	-	-	-	0.265	0.0104	0.265
-	-	-	-	2	-	-	0.0105	0.267
32	-	-	-	-	-	-	0.0108	0.274
-	-	-	-	-	2	0.280	0.0110	0.280
-	-	-	29	-	-	-	0.0113	0.287
-	-	-	-	3	-	-	0.0115	0.292
31	-	-	-	-	-	-	0.0116	0.295
-	33	-	-	-	-	0.300	0.0118	0.300

SWG	W&M	BWG	AWG	EMW (UK)	ASW (US)	Metric R40	in	mm
-	-	30	-	-	3	_	0.0120	0.305
30	-	-	-	-	-	0.315	0.0124	0.315
-	-	-	-	4	-	-	0.0125	0.317
-	-	-	28	-	-	-	0.0126	0.320
-	32	-	-	-	-	-	0.0128	0.325
-	-	29	-	-	4	-	0.0130	0.330
-	31	-	-	-	-	0.335	0.0132	0.335
29	-	-	-	-	-	-	0.0136	0.345
-	30	28	-	-	5	0.355	0.0140	0.355
-	-	-	27	-	-	-	0.0142	0.361
-	-	-	-	5	-	-	0.0145	0.368
28	-	-	-	-	-	0.375	0.0148	0.375
-	29	-	-	6	-	-	0.0150	0.381
-	-	-	-	-	-	0.400	0.0157	0.400
-	-	-	26	-	-	-	0.0159	0.404
-	-	27	-	-	6	-	0.0160	0.406
-	28	-	-	-	-	-	0.0162	0.411
27	-	-	-	-	-	-	0.0164	0.417
-	-	-	-	-	-	0.425	0.0167	0.425
-	27	-	-	-	-	-	0.0173	0.439
-	-	-	-	7	-	-	0.0175	0.444
-	-	-	25	-	-	0.450	0.0177	0.450
-	-	-	-	-	-	-	0.0179	0.455
26	-	26	-	-	7	-	0.0180	0.457
-	26	-	-	-	-	-	0.0181	0.460
-	-	-	-	-	-	0.475	0.0187	0.475
-	-	-	-	8	-	-	0.0190	0.483
-	-	-	-	-	-	0.500	0.0197	0.500
25	-	25	-	-	8	-	0.0200	0.508
-	-	-	24	-	-	-	0.0201	0.511
-	25	-	-	-	-	-	0.0204	0.518
-	-	-	-	-	-	0.530	0.0209	0.530
24	-	24	-	9	9	0.560	0.0220	0.560
-	-	-	23	-	-	-	0.0226	0.574
-	24	-	-	-	-	-	0.0230	0.584
-	-	-	-	-	-	0.600	0.0236	0.600
23	-	-	-	-	10	-	0.0240	0.610
-	-	-	-	10	-	-	0.0245	0.622
-	-	-	-	-	-	0.630	0.0248	0.630

SWG	W&M	BWG	AWG	EMW (UK)	ASW (US)	Metric R40	in	mm
-	-	23	-	-	-	-	0.0250	0.635
-	-	-	22	-	-	-	0.0253	0.643
-	23	-	-	-	-	-	0.0258	0.655
-	-	-	-	-	11	-	0.0260	0.660
-	-	-	-	-	-	0.670	0.0264	0.670
-	-	-	-	11	-	-	0.0270	0.686
22	-	22	-	12	-	0.710	0.0280	0.710
-	-	-	21	-	-	-	0.0285	0.724
-	22	-	-	-	-	-	0.0286	0.726
-	-	-	-	-	12	-	0.0290	0.737
-	-	-	-	-	-	0.750	0.0295	0.750
-	-	-	-	13	-	-	0.0305	0.775
-	-	-	-	-	13	-	0.0310	0.787
-	-	-	-	-	-	0.800	0.0315	0.800
-	21	-	-	-	-	-	0.0317	0.805
21	-	21	20	14	-	-	0.0320	0.813
-	-	-	-	-	14	-	0.0330	0.838
-	-	-	-	-	-	0.850	0.0335	0.850
-	20	-	-	-	-	-	0.0348	0.884
-	-	20	-	15	15	-	0.0350	0.889
-	-	-	-	-	-	0.900	0.0354	0.900
-	-	-	19	-	-	-	0.0359	0.912
20	-	-	-	16	-	-	0.0360	0.914
-	-	-	-	-	16	-	0.0370	0.940
-	-	-	-	-	-	0.950	0.0374	0.950
-	-	-	-	17	-	-	0.0380	0.965
-	-	-	-	-	17	-	0.0390	0.991
-	-	-	-	-	-	1.000	0.0394	1.000
19	-	-	-	18	-	-	0.0400	1.016
-	-	-	18	-	-	-	0.0403	1.024
-	19	-	-	-	18	-	0.0410	1.041
-	-	-	-	-	-	1.060	0.0417	1.060
-	-	19	-	19	-	-	0.0420	1.067
-	-	-	-	20	19	-	0.0430	1.092
-	-	-	-	-	-	1.120	0.0441	1.120
-	-	-	-	21	-	-	0.0445	1.130
-	-	-	-	-	20	-	0.0450	1.143
-	-	-	17	-	-	-	0.0453	1.151
-	-	-	-	22	21	-	0.0470	1.194
-	18	-	-	-	-	-	0.0475	1.206
18	-	-	-	-	-	-	0.0480	1.219

SWG	W&M	BWG	AWG	EMW (UK)	ASW (US)	Metric R40	in	mm
-	-	18	-	23	22	-	0.0490	1.245
-	-	-	-	-	-	1.250	0.0492	1.250
-	-	-	16	-	-	-	0.0508	1.290
-	-	-	-	-	23	-	0.0510	1.295
-	-	-	-	-	-	1.320	0.0520	1.320
-	-	-	-	24	-	-	0.0530	1.346
-	17	-	-	-	-	-	0.0540	1.372
-	-	-	-	-	24	-	0.0550	1.397
-	-	-	-	-	-	1.400	0.0551	1.400
17	-	-	-	25	-	-	0.0560	1.422
-	-	-	15	-	-	-	0.0571	1.450
-	-	17	-	-	-	-	0.0580	1.473
-	-	-	-	-	25	-	0.0590	1.499
-	-	-	-	-	-	1.500	0.0591	1.500
-	-	-	-	26	-	-	0.0605	1.537
-	16	-	-	-	-	-	0.0625	1.587
-	-	-	-	-	26	1.600	0.0630	1.600
-	-	-	-	27	-	-	0.0640	1.626
16	-	-	-	-	-	-	0.0640	1.626
-	-	-	14	-	-	-	0.0641	1.628
-	-	16	-	-	-	-	0.0650	1.651
-	-	-	-	-	-	1.700	0.0669	1.700
-	-	-	-	-	27	-	0.0670	1.702
-	-	-	-	28	-	-	0.0685	1.740
-	-	-	-	-	-	1.800	0.0709	1.800
-	-	-	-	-	28	-	0.0710	1.803
-	-	-	-	29	-	-	0.0715	1.816
15	15	15	13	-	-	-	0.0720	1.829
-	-	-	-	-	-	1.900	0.0748	1.900
-	-	-	-	30	29	-	0.0750	1.905
-	-	-	-	-	-	2.000	0.0787	2.000
14	14	-	-	-	30	-	0.0800	2.032
-	-	-	12	-	-	-	0.0808	2.052
-	-	14	-	-	-	-	0.0830	2.108
-	-	-	-	-	-	2.120	0.0835	2.120
-	-	-	-	-	31	-	0.0850	2.159
-	-	-	-	-	-	2.240	0.0882	2.240
-	-	-	-	-	32	-	0.0900	2.286
-	-	-	11-	-	-	-	0.0907	2.304
-	13	-	-	-	-	-	0.0915	2.324
13	-	-	-	-	-	-	0.0920	2.337
-	-	-	-	-	-	2.360	0.0929	2.360

SWG	W&M	BWG	AWG	EMW (UK)	ASW (US)	Metric R40	in	mm
-	-	13	-	-	33	-	0.0950	2.413
-	-	-	-	-	-	2.500	0.0984	2.500
-	-	-	-	-	34	-	0.1000	2.540
-	-	-	10	-	-	-	0.1019	2.588
12	-	-	-	-	-	-	0.1040	2.642
-	-	-	-	-	-	2.650	0.1043	2.650
-	12	-	-	-	-	-	0.1055	2.680
-	-	-	-	-	35	-	0.1060	2.692
-	-	12	-	-	-	-	0.1090	2.769
-	-	-	-	-	-	2.800	0.1102	2.800
-	-	-	-	-	36	-	0.1120	2.845
-	-	-	9	-	-	-	0.1144	2.906
11	-	-	-	-	-	-	0.1160	2.946
-	-	-	-	-	37	-	0.1180	2.997
-	-	11	-	-		3.000	0.1181	3.000
-	-	-	-	-	-	-	0.1200	3.048
-	11	-	-	-	-	-	0.1205	3.061
-	-	-	-	-	-	3.150	0.1240	3.150
-	-	-	-	-	38	-	0.1250	3.175
10	-	-	-	-	-	-	0.1280	3.251
-	-	-	-	-	39	-	0.1300	3.302
-	-	-	-	-	-	3.350	0.1319	3.350
-	-	10	-	-	-	-	0.1340	3.404
-	10	-	-	-	-	-	0.1350	3.429
-	-	-	-	-	40	-	0.1380	3.505
-	-	-	-	-	-	3.550	0.1398	3.550
9	-	-	-	-	-	-	0.1440	3.658
-	-	-	-	-	41	-	0.1460	3.708
-	-	-	-	-	-	3.750	0.1476	3.750
-	-	9	-	-	-	-	0.1480	3.759
-	9	-	-	-	-	-	0.1483	3.767
-	-	-	-	-	42		0.1540	3.912
-	-	-	-	-	-	4.000	0.1575	4.000

SHEET METAL GAUGES

SYSTEMS

Like wire sizes, sheet metals have been made using a number of standard gauges, the standard used frequently depending on the type of material. In the UK, most manufacturers, but not all, have conformed to one of two standards: Standard Wire Gauge (SWG) and Birmingham Gauge (BG).

SWG sizes for sheet metal are the same as those for SWG wire gauges. The values for BG are given in the table below and should not be confused with Birmingham Wire Gauge, which is different. In the US, sheet steel is based on the Manufacturer's Standard Gauge for Steel (MSG), given below. The comments on page 133 regarding metrication also apply here.

Table 22.2. Sheet metal gauges: comparison

SWG	BG	MSG (US)	Metric R40	in	mm
-	-	36	-	0.00670	0.170
37	-	-	-	0.00680	0.173
-	35	-	-	0.00690	0.175
-	-	-	0.180	0.00709	0.180
-	-	-	0.190	0.00748	0.190
-	-	35	-	0.00750	0.190
36	-	-	-	0.00760	0.193
-	34	-	-	0.00770	0.196
-	-	-	0.200	0.00787	0.200
-	-	34	-	0.00820	0.208
-	-	-	0.212	0.00835	0.212
35	-	-	-	0.00840	0.213
-	33	-	-	0.00870	0.221
-	-		0.224	0.00882	0.224
-	-	33	-	0.00900	0.229
34	-	-	-	0.00920	0.234
-	-	-	0.236	0.00929	0.236
-	-	32	-	0.00970	0.246
-	32	-	-	0.00980	0.249
-	-	-	0.250	0.00984	0.250

SWG	BG	MSG (US)	Metric R40	in	mm
33	-	-	-	0.01000	0.254
-	-	-	0.265	0.01043	0.265
-	-	31	-	0.01050	0.267
32	-	-	-	0.01080	0.274
-	31	-	-	0.01100	0.279
-	-	-	0.280	0.01102	0.280
31	-	-	-	0.01160	0.295
-	-	-	0.300	0.01181	0.300
-	30	-	-	0.01200	0.305
-	30	-	-	0.01230	0.312
30	-	-	0.315	0.01240	0.315
-	-	-	0.335	0.01319	0.335
-	29	-	-	0.01350	0.343
29	-	-	-	0.01360	0.345
-	29	-	-	0.01390	0.353
-	-	-	0.355	0.01400	0.355
28	-	-	0.375	0.01476	0.375
-	-	28	-	0.01490	0.378
-	28	-	-	0.01562	0.397
-	-	-	0.400	0.01575	0.400
27	-	27	-	0.01640	0.417
-	-	-	0.425	0.01673	0.425
-	27	-	-	0.01745	0.443
-	-	-	0.450	0.01772	0.450
-	-	26	-	0.01790	0.455
26	-	-	-	0.01800	0.457
-	-	-	0.475	0.01870	0.475
-	26	-	-	0.01961	0.498
-	-	-	0.500	0.01969	0.500
25	-	-	-	0.02000	0.508
-	-	-	0.530	0.02087	0.530
-	-	25	-	0.02090	0.531
-	25	-	-	0.02204	0.560
24	-	-	0.560	0.02205	0.560
-	-	-	0.600	0.02362	0.600
-	-	24	-	0.02390	0.607
23	-	-	-	0.02400	0.610
-	24	-	-	0.02476	0.629

SWG	BG	MSG (US)	Metric R40	in	mm
-	-	-	0.630	0.02480	0.630
-	-	-	0.670	0.02638	0.670
-	-	23	-	0.02690	0.683
-	23	-	-	0.02782	0.707
22	-	-	0.710	0.02800	0.710
-	-	-	0.750	0.02953	0.750
-	-	22	-	0.02990	0.759
-	22	-	-	0.03125	0.794
-	-	-	0.800	0.03150	0.800
21	-	-	-	0.03200	0.813
-	-	21	-	0.03290	0.836
-	-	-	0.850	0.03346	0.850
-	21	-	-	0.03490	0.886
-	-	-	0.900	0.03543	0.900
-	-	20	-	0.03590	0.912
20	-	-	-	0.03600	0.914
-	-	-	0.950	0.03740	0.950
-	20	-	-	0.03920	0.996
-	-	-	1.000	0.03937	1.000
19	-	-	-	0.04000	1.016
-	-	-	1.060	0.04173	1.060
-	-	19	-	0.04180	1.062
-	19	-	-	0.04400	1.118
-	-	-	1.120	0.04409	1.120
-	-	18	-	0.04780	1.214
18	-	-	-	0.04800	1.219
-	-	-	1.250	0.04921	1.250
-	18	-	-	0.04950	1.257
-	-	-	1.320	0.05197	1.320
-	-	17	-	0.05380	1.367
-	-	-	1.400	0.05512	1.400
-	17	-	-	0.05560	1.412
17	-	-	-	0.05600	1.422
-	-	-	1.500	0.05906	1.500
-	-	16	-	0.05980	1.519
-	16	-	-	0.06250	1.587
-	-	-	1.600	0.06299	1.600
16	-	-	-	0.06400	1.626

SWG	BG	MSG (US)	Metric R40	in	mm
-	-	-	1.700	0.06693	1.700
-	-	15	-	0.06730	1.709
-	15	-	-	0.06990	1.775
-	-	-	1.800	0.07087	1.800
15	-	-	-	0.07200	1.829
-	-	14	-	0.07470	1.897
-	-	-	1.900	0.07480	1.900
-	14	-	-	0.07850	1.994
-	-	-	2.000	0.07874	2.000
14	-	-	-	0.08000	2.032
-	-	-	2.120	0.08346	2.120
-	-	-	2.240	0.08819	2.240
-	13	-	-	0.08820	2.240
-	-	13	-	0.08970	2.278
13	-	-	-	0.09200	2.337
-	-	-	2.360	0.09291	2.360
-	-	-	2.500	0.09843	2.500
-	12	-	-	0.09910	2.517
12	-	-	-	0.10400	2.642
-	-	-	2.650	0.10433	2.650
-	-	12	-	0.10460	2.657
-	-	-	2.800	0.11024	2.800
-	11	-	-	0.11130	2.827
11	-	-	-	0.11600	2.946
-	-	-	3.000	0.11811	3.000
-	-	11	-	0.11960	3.038
-	-	-	3.150	0.12402	3.150
-	10	-	-	0.12500	3.175
10	-	-	-	0.12800	3.251
-	-	10	-	0.13450	3.416

CHAPTER 23
MATERIAL SPECIFICATIONS

STEELS

STANDARDS AND SPECIFICATIONS

Reference to British Standard BS970 1983/1991 will show how impracticable it would be to cover this subject in full. Even in industry, only a few per cent of the steels specified are in common use. The data below lists the types of steels most commonly used in the small workshop and stocked by major suppliers, as well as their nearest equivalents from BS970 1955 (EN numbers) and AISI/SAE specifications. As the steels listed are not identical, it is impossible to give exact equivalents but even so those given should be close enough for most applications. However, for critical applications further investigation is advised.

Availability

Most of the specifications are now found in international and world standards, which use the same references, e.g. 230M07. The fact that a steel is quoted below does not necessarily mean that it is easily obtainable in the quantities likely to be required by the small-workshop user.

Coding system

The specification consists of three digits, a letter and a further two digits, although other letters may be added, e.g. to indicate the addition of lead (Pb). For example, the free-cutting steel, formerly known as 'EN1A leaded', has a similar steel in the later standard but with the number 230M07 Pb.

The three digits indicate the percentage of additives, manganese, etc. It is sufficient to know the basic material grouping, i.e.:

000–199	Carbon steels containing manganese
200–299	Free-cutting versions of the above
300–499	Stainless and valve steels
500–999	Alloy steels other than stainless.

The letter indicates the basis on which the standard is established, e.g:

M	Mechanical characteristics
A	Analysis (chemical)
H	Hardening, specific properties
S	Stainless

M is by far the most commonly used letter.

For carbon steels the last two digits show the carbon content, e.g.:

07	0.07%
30	0.3%
40	0.40%.

In cases where the content is 1% or more the last two digits are 99.

Bar markings

Steel supplied in full lengths is colored at one end to indicate its type. This is a supplier's marking and varies from one supplier to another. It cannot therefore be relied upon unless you are familiar with the supplier and their color coding standard.

CARBON CONTENT

The main effect of carbon is to govern the degree of heat treatment that is practicable. Steels with a carbon content of less than 0.3% are frequently referred to as 'mild' steels and are used where strength and hardness are less critical.

Steels containing more than 0.3% carbon are intended for heat treatment, which is carried out by the supplier and/or the user. At the lower values, the steel is primarily used where toughness, rather than hardness, is the important factor, e.g. high tensile bolts, springs, etc.

Steels with a carbon content in excess of 0.7% can be hardened to a higher degree and are used where hardness, rather than toughness, is the requirement. Very few steels above 0.7% are listed in BS970; tool steels, silver steel (drill rod), etc. are covered by other standards.

BENDING QUALITY

Most steels, as supplied, cannot be bent without using a large internal radius. Therefore, bending quality steel should be used where 90° bends with an average internal radius, e.g. equal to material thickness, are to be made. Other steels can be bent if they are tempered by raising their temperature to red heat and cooling slowly.

Table 23.1. Comparison of steels

BS970 1991	BS970 1955	AISI/SAE	MC
040A10	EN2A	1010	
070M20	EN3B	1020	F
070M55	EN9	1055	P
080A15	EN3B	1016	
080A42	EN8D	1042	
080M15	EN32C	1016	F
080M30	EN6	1030	
080M30	EN5	1030	
080M40	EN8	1040	F
080M50	EN43A	1049	
150M19	EN14A	1024	
150M36	EN15	-	F
210M15	EN32M	1117	G
212A42	EN8DM	-	G
212M36	EN8M	1140	G
212M44	EN8M	-	G
214M15	EN202	1118	E
216M36	EN15AM	1137	G
220M07	EN1A	1113	E
226M44	EN8M	1144	G
230M07	EN1A	1213	E
230M07Pb	EN1APb	12L14	E+
605M36	EN16	-	P
606M36	EN16M	-	P
635M15	EN351	-	-
655M13	EN36	-	-
665M17	EN34	4615	-
708M40	EN19	4140	P
722M24	EN40B	-	-
805M20	EN362	8620	-
817M40	EN24	4340	P
835M30	EN30B	-	-

MC	ease of machining, where available
E	excellent
F	fair
G	good
P	poor

USES

Many steels have been established with a particular type of application in mind, e.g. free cutting for mass production, components requiring case hardening, or welding. However, generally speaking, most steels can be used for more than one type of application. The following guidelines give the requirements for particular applications.

Case hardening

Most steels can be case hardened, but a part that requires a hard surface may also be subject to high stress and its core must be tough enough to withstand this. If the wrong steel is chosen, the hardened surface may crack away because of movement in the core due to the stresses placed on it. Readers should be aware of this although it will rarely be a problem in the small workshop because any mild steel (steel with less than 0.3% carbon) is likely to give satisfactory results. However, 080M15 is considered particularly suitable for lightly stressed components.

Welding

Non-free-cutting carbon steels with a carbon content of less than 0.3% are generally suitable for welding. However, free-cutting carbon steels are not ideal because their sulfur content produces many blow holes, although satisfactory results may be obtained using special electrodes. Some alloy steels can be welded but, again, special electrodes and techniques may be required.

Heat treatment

Steels with a carbon content in excess of 0.3% are suitable for heat treatment, either for tensile strength or for hardness. This is a complex subject, but suppliers will provide data for heat treating particular steels, as they also do for silver steels, gauge plate, etc.

Shortlist

The following is a shortlist of steels which are suited to the small workshop and are most likely to be easily obtainable.

Lightly stressed engineering parts:
Free cutting	220/230M07
Adequate machining	070M20

Lightly stressed case-hardened parts:
Ideal	080M15
Adequate	070M20

Welded parts:
Adequate	070M20

Good all-round steel for lightly stressed parts:
Adequate	070M20

Heat treatable for medium-stressed parts:
Free cutting:	212M36
	212A42
	212M44
Adequate machining	080M40

STAINLESS AND VALVE STEELS

BS970 also quotes stainless steels using the same coding system, but with the letter S after the first three digits. The last two digits are of minimal significance, so some suppliers refer to these steels by the first three digits only, e.g. 303, 306, etc. A former method of specifying stainless, sometimes still used, was to indicate the chromium and nickel content, e.g. 18.8 stainless contained 18% chrome and 8% nickel.

CHROMIUM CONTENT

The essential content of stainless steel is chromium in quantities of 10.5% or more, rendering the steel highly resistant to corrosion. While some alloy steels (500 plus numbers) have a chromium content they are not considered stainless as the amount is less than 10.5%.

Some stainless steels are suited to very high temperatures. These are called valve steels and are also included in the 300/400 numbers.

TYPES

Stainless steels are divided into three groups: Austenitic, Ferretic and Martensitic. Martensitic is sometimes shown as a subdivision of Ferretic.

Corrosion resistance is not total and different grades will respond to various chemicals by different amounts. As a general rule, Austenitic steels have the better corrosion resistance but cannot be hardened, while Martensitic steels have a lower corrosion resistance but can be hardened. Ferretic steels have good corrosion resistance and cannot be hardened; they also have good ductility and are mainly available in sheet and strip form.

Table 23.2. Comparison of stainless steels

BS970 1983	BS970 1955	AISI/ SAE	MC	Type
303S31	EN58M	303	G	Aust
303S42	EN58AM	303Se	G	Aust
304S15	EN58E	304	F	Aust
316S31	EN58J	316	F	Aust
321S31	EN58B	321	F	Aust
326S36	EN58JM	-	G	Aust
416S41	EN56AM	416Se	G	Mart
441S49	EN57	-	G	Mart

USES

Machining

The grades shown in the table assume that the necessary precautions are taken when machining stainless. Surprisingly, while Austenitic grades cannot be hardened, they do work-harden. Using lower speeds with a higher feed rate and a coolant will considerably reduce this tendency, as will using tools with lower rake and relief angles.

Shortlist

Type 303 is likely to prove to be a good choice for most small workshop applications where heat treatment is not required. It is not ideal, though not impossible, to weld this material; types 304, 316 and 321 are preferred in this situation.

MATERIAL WEIGHTS

Precise weights for materials depend on the grade of material, e.g. 70/30 Brass or 60/40 Brass. However, in practice these variations result in only small differences. The following tables although not precise, should be accurate enough for most situations where material weights are required.

In the case of woods, there can be large differences in weight, and a few types are included for guidance. The weights given are for air-dried timber, but green woods can be up to 50% heavier. Even for a given species, weights can vary by as much as 20%. Some exotic woods, such as South African leadwood, used for turning, are up to 60% heavier than oak and a few have a specific gravity greater than 1, meaning that they sink rather than float in water.

Table 23.3. Material weights per cubic centimeter

Material	Weight	
	kg	lb
Aluminum	0.00280	0.0062
Brass	0.00840	0.0185
Bronze	0.00890	0.0196
Cast iron	0.00730	0.0161
Copper	0.00894	0.0197
Glass	0.00250	0.0055
Lead	0.01100	0.0243
Mercury	0.01354	0.0299
Nickel silver	0.00861	0.0190
Nylon	0.00114	0.0025
Polyethylene	0.00093	0.0021
PVC	0.00147	0.0032
Stainless steel	0.00790	0.0174
Steel	0.00786	0.0173
Ash	0.00066	0.0014
Beech	0.00072	0.0016
Cedar	0.00053	0.0012
Oak	0.00075	0.0017
Pine, white	0.00048	0.0011

Table 23.4. Material weights per cubic inch

Material	Weight	
	kg	lb
Aluminum	0.0459	0.1012
Brass	0.1377	0.3035
Bronze	0.1458	0.3215
Cast iron	0.1196	0.2637
Copper	0.1465	0.3230
Glass	0.0410	0.0903
Lead	0.1803	0.3974
Mercury	0.2219	0.4892
Nickel silver	0.1411	0.3111
Nylon	0.0187	0.0412
Polyethylene	0.0152	0.0336
PVC	0.0240	0.0530
Stainless steel	0.1295	0.2854
Steel	0.1288	0.2840
Ash	0.0108	0.0237
Beech	0.0118	0.0260
Cedar	0.0087	0.0191
Oak	0.0123	0.0272
Pine, white	0.0079	0.0174

CHAPTER 24

MISCELLANEOUS DATA

LARGE NUMBERS

Million	UK & US	106	1,000,000	one thousand × one thousand
Billion	UK	10^{12}	1,000,000,000,000	one million × one million
	US	10^9	1,000,000,000	one thousand x 1 million
Trillion	UK	10^{18}	1,000,000,000,000,000,000	one million × one billion
	US	10^{12}	1,000,000,000,000	one million × 1 million

ORDERS OF MAGNITUDE

Multiplying factor		Prefix	Symbol
1,000,000,000,000	$= 10^{12}$	tera	T
1,000,000,000	$= 10^9$	giga	G
1,000,000	$= 10^6$	mega	M
1,000	$= 10^3$	kilo	k
100	$= 10^2$	hecto*	h*
10	$= 10^1$	deca*	da*
1	$= 10^0$	-	
0.1	$= 10^{-1}$	deci*	d*
0.01	$= 10^{-2}$	centi*	c*
0.001	$= 10^{-3}$	milli	m
0.000,001	$= 10^{-6}$	micro	μ
0.000,000,001	$= 10^{-9}$	nano	n
0.000,000,000,001	$= 10^{-12}$	pico	p

Non-preferred term.

TYPICAL USAGE

Measurements of length

The unit of length is the meter, with the longest unit being the kilometer (km) and the shorter units being the millimeter (mm) and micron (μm, or 0.001 millimeter). The use of the term 'micron' in place of micrometer is to avoid confusion with the measuring instrument. The centimeter is still used in some industries but is definitely non-preferred for engineering measurements.

Electrical units

In electrical work, the prefixes mega-, kilo- and milli- are frequent found in relation to voltage(volts, V), current (amps, A), power (watts, W) and resistance (ohms, Ω) while the prefixes micro-, nano- and pico- are used in relation to capacitance, which is measured in farads (F). Inductance is normally given in millihenries (mH).

In all these units the larger and smaller multipliers can be used if necessary.

ROMAN NUMERALS

I	=	1	XI	=	11	XXX	=	30
II	=	2	XII	=	12	XL	=	40
III	=	3	XIII	=	13	L	=	50
IV	=	4	X1V	=	14	LX	=	60
V	=	5	XV	=	15	LXX	=	70
VI	=	6	XVI	=	16	LXXX	=	80
VII	=	7	XVII	=	17	XC	=	90
VIII	=	8	XVIII	=	18	C	=	100
IX	=	9	XIX	=	19	D	=	500
X	=	10	XX	=	20	M	=	1000

PAPER SIZES

A1	594 × 841
A2	420 × 594
A3	297 × 420
A4	210 × 297
A5	148 × 210
A6	105 × 148

ISO PREFERRED NUMBERS

The following lists the ISO series of preferred numbers around which standards are intended to be established. This is arranged to give a consistent percentage increase between numbers but with some rounding of the values to give more usable numbers for practical applications. Further rounding may take place when applied to a particular application. The mathematical background to the numbers is the series comprising $10^{n/10}$, $10^{n/20}$, $10^{n/40}$ for the R10, R20 and R40 series respectively.

As an example, for the R20 series:

$10^{1/20}$	=	1.1220 rounded	=	1.12
$10^{2/20}$	=	1.2589 rounded	=	1.25
$10^{3/20}$	=	1.4125 rounded	=	1.40
$10^{4/20}$	=	1.5849 rounded	=	1.60
$10^{5/20}$	=	1.7783 rounded	=	1.80
up to				
$10^{20/20}$	=	10	=	10.00

The table below lists full details for the R40 series, in which every second row of numbers is included in the R20 series, every fourth in the R10 series, and every eighth in the R5 series. An R80 series is also defined within the standards.

R5	R10	R20	Hundredths	Tenths	Units	Tens
I	I	I	-	0.100	1.00	10.0
			-	0.106	1.06	10.6
		I	-	0.112	1.12	11.2
			-	0.118	1.18	11.8
	I	I	-	0.125	1.25	12.5
			-	0.132	1.32	13.2
		I	-	0.140	1.40	14.0
			-	0.150	1.50	15.0
I	I	I	-	0.160	1.60	16.0
			-	0.170	1.70	17.0
		I	-	0.180	1.80	18.0
			-	0.190	1.90	19.0
	I	I	0.020	0.200	2.00	20.0
			0.021	0.212	2.12	21.2
		I	0.022	0.224	2.24	22.4
			0.024	0.236	2.36	23.6
I	I	I	0.025	0.250	2.50	25.0
			0.026	0.265	2.65	-
		I	0.028	0.280	2.80	-
			0.030	0.300	3.00	-
	I	I	0.032	0.315	3.15	-
			0.034	0.335	3.35	-
		I	0.036	0.355	3.55	-

R5	R10	R20	Hundredths	Tenths	Units	Tens
			0.038	0.375	3.75	-
I	I	I	0.040	0.400	4.00	-
			0.042	0.425	4.25	-
		I	0.045	0.450	4.50	-
			0.048	0.475	4.75	-
	I	I	0.050	0.500	5.00	-
			0.053	0.530	5.30	-
		I	0.056	0.560	5.60	-
			0.060	0.600	6.00	-
I	I	I	0.063	0.630	6.30	-
			0.067	0.670	6.70	-
		I	0.071	0.710	7.10	-
			0.075	0.750	7.50	-
	I	I	0.080	0.800	8.00	-
			0.085	0.850	8.50	-
		I	0.090	0.900	9.00	-
			0.095	0.950	9.50	-

I = included.

DOMESTIC OVEN TEMPERATURES

Anyone thinking of using a domestic oven for workshop activities should give due consideration to its main purpose: the preparation of food. Because of this, any activity that may contaminate the oven should not be attempted. However, processes such as tempering clean tool steel should be quite acceptable.

The following table gives an indication of the temperature equivalents but it is advisable to consult the manufacturer's manual that comes with the oven or to use a reliable oven thermometer.

Regulo no.	Temperature	
	°F	°C
1/4	225	107
1/2	250	121

1	275	135
2	300	149
3	325	163
4	350	177
5	375	191
6	400	204
7	425	218
8	450	232
9	475	246

THE ELEMENTS

Details of the content of a material are frequently given in terms of the percentages of the different elements that make up the material. This often quotes their symbols rather than their full name, which is not always easy to determine from the symbol itself, e.g. Ag for silver.

Knowing the detailed content of home workshop materials is rarely of great importance, but it can add considerably to the interest, which is why the following list of elements and their symbols is included.

Note that the melting points given for the more common metals are for the pure metal and these will vary if the metal is alloyed with other materials. The letter Y (yes) or N (no) indicates whether the element is metallic.

For further information about the physical properties of the elements, such as density, specific heat etc., a suitable reference book should be consulted.

Symbol	Element	Metallic	Melting point (°C)
Ac	Actinium	N	
Ag	Silver	Y	962
Al	Aluminum	Y	660
Am	Americium	N	
Ar	Argon	N	

Symbol	Element	Metallic	Melting point (°C)
As	Arsenic	N	
At	Astatine	N	
Au	Gold	Y	1064
B	Boron	N	
Ba	Barium	Y	
Be	Beryllium	Y	1285
Bi	Bismuth	Y	
Bk	Berkelium	N	
Br	Bromine	N	
C	Carbon	N	
Ca	Calcium	Y	
Cd	Cadmium	Y	321
Ce	Cerium	Y	
Cf	Californium	N	
Cl	Chlorine	N	
Cm	Curium	N	
Co	Cobalt	Y	
Cr	Chromium	Y	1860
Cs	Caesium	Y	
Cu	Copper	Y	1084
Dy	Dysprosium	N	
Er	Erbium	N	
Es	Einsteinium	N	
Eu	Europium	N	
F	Fluorine	N	
Fe	Iron	Y	1540
Fm	Fermium	N	
Fr	Francium	N	
Ga	Gallium	Y	
Gd	Gadolinium	N	
Ge	Germanium	Y	
H	Hydrogen	N	
He	Helium	N	
Hf	Hafnium	Y	

Symbol	Element	Metallic	Melting point (°C)
Hg	Mercury	Y	
Ho	Holmium	N	
I	Iodine	N	
In	Indium	Y	
Ir	Iridium	Y	
K	Potassium	Y	
Kr	Krypton	N	
La	Lanthanum	N	
Li	Lithium	Y	
Lu	Lutecium	N	
Lw	Lawrencium	N	
Md	Mendelevium	N	
Mg	Magnesium	Y	
Mn	Manganese	Y	
Mo	Molybdenum	Y	
N	Nitrogen	N	
Na	Sodium	Y	
Nb	Niobium	Y	
Nd	Neodymium	N	
Ne	Neon	N	
Ni	Nickel	Y	1455
No	Nobelium	N	
Np	Neptunium	N	
O	Oxygen	N	
Os	Osmium	Y	
P	Phosphorus	N	
Pa	Protactinium	N	
Pb	Lead	Y	327
Pd	Palladium	Y	
Pm	Promethium	N	
Po	Polonium	N	
Pr	Praseodymium	N	
Pt	Platinum	Y	1772
Pu	Plutonium	N	

Symbol	Element	Metallic	Melting point (°C)
Ra	Radium	N	
Rb	Rubidium	Y	
Re	Rhenium	Y	
Rh	Rhodium	Y	
Rn	Radon	N	
Ru	Ruthenium	Y	
S	Sulfur	N	
Sb	Antimony	Y	
Sc	Scandium	N	
Se	Selenium	N	
Si	Silicon	Y	
Sm	Samarium	N	
Sn	Tin	Y	232
Sr	Strontium	Y	
Ta	Tantalum	Y	
Tb	Terbium	N	
Tc	Technetium	N	
Te	Tellurium	Y	
Th	Thorium	Y	
Ti	Titanium	Y	
Tl	Thallium	Y	
Tm	Thulium	N	
U	Uranium	Y	
V	Vanadium	Y	
W	Tungsten	Y	3387
Xe	Xenon	N	
Y	Yttrium	N	
Yb	Ytterbium	N	
Zn	Zinc	Y	420
Zr	Zirconium	Y	

MACHINE TOOL AXES LETTER REFERENCES

AXES REFERENCES

There is only a limited need to refer to machine tool motions by the established letter references. However, some authors refer to machine tool motions in this way and, because of this, it is necessary to understand their meanings.

Letter references have been standardized for all forms of machine tools but only the vertical milling machine and the center lathe are found in the majority of workshops (see drawings). Both drawings show the references for the three axes, although in the case of the lathe, the Y axis is rarely applicable.

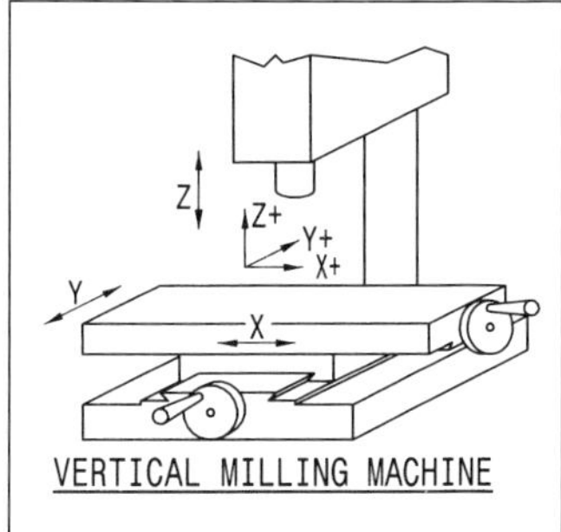

24-01: Vertical milling machine

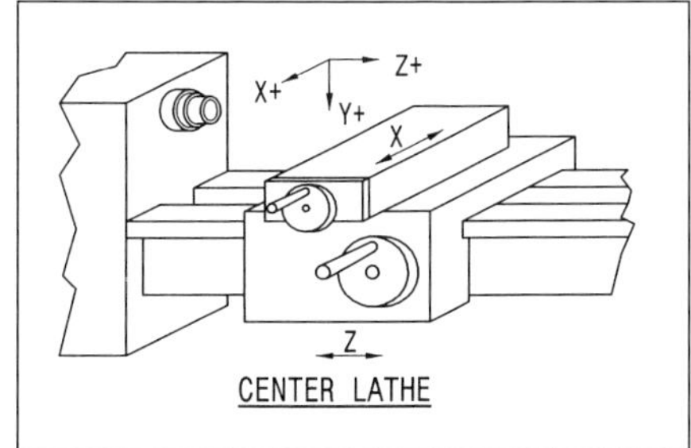

24-02: Center lathe

POSITIVE/NEGATIVE DIRECTIONS.

The need to define the polarity of each axis results from their use in Computer Numerical Control (CNC) and, in this case, the point to note is that CNC programming works on the basis of the relative cutter/workpiece direction, irrespective of whether the situation is workpiece moving/cutter stationary or cutter moving/workpiece stationary. Typically, therefore, in the case of the X axis on the milling machine, positive movement occurs when the cutter moves to the right of the workpiece, i.e. when the table moves to the left.

Positive movement of the Z axis (cutter spindle) occurs when the cutter is withdrawn from the workpiece.

EQUIPMENT PROTECTION

Enclosures into which equipment is fitted are frequently, although not always, electrical. It is therefore necessary to provide two basic forms of protection:

- to protect the operator from dangers inside
- to protect the internal equipment from harmful external items, e.g. dust and water.

However, equipment often needs to be ventilated and openings have to be provided to achieve this.

IP (INGRESS PROTECTION) NUMBERS

The protection provided is indicated by a two-digit number with the prefix IP. The first digit indicates protection from solid objects, and the second digit indicates protection from liquids.

First digit: protection against the entry of solid objects, i.e.:

1. A 50mm sphere
2. A 12.5mm bar
3. A 2.5mm bar
4. A 1mm bar
5. Dust
6. Totally dust proof.

In practical terms, 1 provides protection for the entire hand and 2 for a finger, while 3 protects against the entry of a screw driver and 4 a piece of wire. Live or moving items inside must be well clear of any openings to maximize the protection given.

Second digit: protection against the entry of liquids, i.e.:

1. Vertical drips
2. Vertical drips, enclosure tilted 15°
3. Spraying at a downward angle of 60°
4. Splashing, from any angle
5. Jets
6. Power jets
7. Temporary immersion
8. Continuous immersion.

A zero in either the first- or second position indicates that there is no protection. The above is a simplified explanation of the standard. Details for carrying out the tests are also laid down.

CHAPTER 25

MATHEMATICAL TABLES

Table 25.1 Sines

°	0	6'	12'	18'	24'	30'	36'	42'	48'	54'
	0	0.1	0.2	0.3	0.4	0.5	0.6	0.7	0.8	0.9
0	0	0.0017	0.0035	0.0052	0.0070	0.0087	0.0105	0.0122	0.0140	0.0157
1	0.0175	0.0192	0.0209	0.0227	0.0244	0.0262	0.0279	0.0297	0.0314	0.0332
2	0.0349	0.0366	0.0384	0.0401	0.0419	0.0436	0.0454	0.0471	0.0488	0.0506
3	0.0523	0.0541	0.0558	0.0576	0.0593	0.0610	0.0628	0.0645	0.0663	0.0680
4	0.0698	0.0715	0.0732	0.0750	0.0767	0.0785	0.0802	0.0819	0.0837	0.0854
5	0.0872	0.0889	0.0906	0.0924	0.0941	0.0958	0.0976	0.0993	0.1011	0.1028
6	0.1045	0.1063	0.1080	0.1097	0.1115	0.1132	0.1149	0.1167	0.1184	0.1201
7	0.1219	0.1236	0.1253	0.1271	0.1288	0.1305	0.1323	0.1340	0.1357	0.1374
8	0.1392	0.1409	0.1426	0.1444	0.1461	0.1478	0.1495	0.1513	0.1530	0.1547
9	0.1564	0.1582	0.1599	0.1616	0.1633	0.1650	0.1668	0.1685	0.1702	0.1719
10	0.1736	0.1754	0.1771	0.1788	0.1805	0.1822	0.1840	0.1857	0.1874	0.1891
11	0.1908	0.1925	0.1942	0.1959	0.1977	0.1994	0.2011	0.2028	0.2045	0.2062
12	0.2079	0.2096	0.2113	0.2130	0.2147	0.2164	0.2181	0.2198	0.2215	0.2233
13	0.2250	0.2267	0.2284	0.2300	0.2317	0.2334	0.2351	0.2368	0.2385	0.2402
14	0.2419	0.2436	0.2453	0.2470	0.2487	0.2504	0.2521	0.2538	0.2554	0.2571
15	0.2588	0.2605	0.2622	0.2639	0.2656	0.2672	0.2689	0.2706	0.2723	0.2740
16	0.2756	0.2773	0.2790	0.2807	0.2823	0.2840	0.2857	0.2874	0.2890	0.2907
17	0.2924	0.2940	0.2957	0.2974	0.2990	0.3007	0.3024	0.3040	0.3057	0.3074
18	0.3090	0.3107	0.3123	0.3140	0.3156	0.3173	0.3190	0.3206	0.3223	0.3239
19	0.3256	0.3272	0.3289	0.3305	0.3322	0.3338	0.3355	0.3371	0.3387	0.3404
20	0.3420	0.3437	0.3453	0.3469	0.3486	0.3502	0.3518	0.3535	0.3551	0.3567
21	0.3584	0.3600	0.3616	0.3633	0.3649	0.3665	0.3681	0.3697	0.3714	0.3730
22	0.3746	0.3762	0.3778	0.3795	0.3811	0.3827	0.3843	0.3859	0.3875	0.3891
23	0.3907	0.3923	0.3939	0.3955	0.3971	0.3987	0.4003	0.4019	0.4035	0.4051
24	0.4067	0.4083	0.4099	0.4115	0.4131	0.4147	0.4163	0.4179	0.4195	0.4210
25	0.4226	0.4242	0.4258	0.4274	0.4289	0.4305	0.4321	0.4337	0.4352	0.4368
26	0.4384	0.4399	0.4415	0.4431	0.4446	0.4462	0.4478	0.4493	0.4509	0.4524
27	0.4540	0.4555	0.4571	0.4586	0.4602	0.4617	0.4633	0.4648	0.4664	0.4679

°	0	6'	12'	18'	24'	30'	36'	42'	48'	54'
	0	0.1	0.2	0.3	0.4	0.5	0.6	0.7	0.8	0.9
28	0.4695	0.4710	0.4726	0.4741	0.4756	0.4772	0.4787	0.4802	0.4818	0.4833
29	0.4848	0.4863	0.4879	0.4894	0.4909	0.4924	0.4939	0.4955	0.4970	0.4985
30	0.5000	0.5015	0.5030	0.5045	0.5060	0.5075	0.5090	0.5105	0.5120	0.5135
31	0.5150	0.5165	0.5180	0.5195	0.5210	0.5225	0.5240	0.5255	0.5270	0.5284
32	0.5299	0.5314	0.5329	0.5344	0.5358	0.5373	0.5388	0.5402	0.5417	0.5432
33	0.5446	0.5461	0.5476	0.5490	0.5505	0.5519	0.5534	0.5548	0.5563	0.5577
34	0.5592	0.5606	0.5621	0.5635	0.5650	0.5664	0.5678	0.5693	0.5707	0.5721
35	0.5736	0.5750	0.5764	0.5779	0.5793	0.5807	0.5821	0.5835	0.5850	0.5864
36	0.5878	0.5892	0.5906	0.5920	0.5934	0.5948	0.5962	0.5976	0.5990	0.6004
37	0.6018	0.6032	0.6046	0.6060	0.6074	0.6088	0.6101	0.6115	0.6129	0.6143
38	0.6157	0.6170	0.6184	0.6198	0.6211	0.6225	0.6239	0.6252	0.6266	0.6280
39	0.6293	0.6307	0.6320	0.6334	0.6347	0.6361	0.6374	0.6388	0.6401	0.6414
40	0.6428	0.6441	0.6455	0.6468	0.6481	0.6494	0.6508	0.6521	0.6534	0.6547
41	0.6561	0.6574	0.6587	0.6600	0.6613	0.6626	0.6639	0.6652	0.6665	0.6678
42	0.6691	0.6704	0.6717	0.6730	0.6743	0.6756	0.6769	0.6782	0.6794	0.6807
43	0.6820	0.6833	0.6845	0.6858	0.6871	0.6884	0.6896	0.6909	0.6921	0.6934
44	0.6947	0.6959	0.6972	0.6984	0.6997	0.7009	0.7022	0.7034	0.7046	0.7059
45	0.7071	0.7083	0.7096	0.7108	0.7120	0.7133	0.7145	0.7157	0.7169	0.7181
46	0.7193	0.7206	0.7218	0.7230	0.7242	0.7254	0.7266	0.7278	0.7290	0.7302
47	0.7314	0.7325	0.7337	0.7349	0.7361	0.7373	0.7385	0.7396	0.7408	0.7420
48	0.7431	0.7443	0.7455	0.7466	0.7478	0.7490	0.7501	0.7513	0.7524	0.7536
49	0.7547	0.7559	0.7570	0.7581	0.7593	0.7604	0.7615	0.7627	0.7638	0.7649
50	0.7660	0.7672	0.7683	0.7694	0.7705	0.7716	0.7727	0.7738	0.7749	0.7760
51	0.7771	0.7782	0.7793	0.7804	0.7815	0.7826	0.7837	0.7848	0.7859	0.7869
52	0.7880	0.7891	0.7902	0.7912	0.7923	0.7934	0.7944	0.7955	0.7965	0.7976
53	0.7986	0.7997	0.8007	0.8018	0.8028	0.8039	0.8049	0.8059	0.8070	0.8080
54	0.8090	0.8100	0.8111	0.8121	0.8131	0.8141	0.8151	0.8161	0.8171	0.8181
55	0.8192	0.8202	0.8211	0.8221	0.8231	0.8241	0.8251	0.8261	0.8271	0.8281
56	0.8290	0.8300	0.8310	0.8320	0.8329	0.8339	0.8348	0.8358	0.8368	0.8377
57	0.8387	0.8396	0.8406	0.8415	0.8425	0.8434	0.8443	0.8453	0.8462	0.8471
58	0.8480	0.8490	0.8499	0.8508	0.8517	0.8526	0.8536	0.8545	0.8554	0.8563
59	0.8572	0.8581	0.8590	0.8599	0.8607	0.8616	0.8625	0.8634	0.8643	0.8652
	0	6'	12'	18'	24'	30'	36'	42'	48'	54'
60	0.8660	0.8669	0.8678	0.8686	0.8695	0.8704	0.8712	0.8721	0.8729	0.8738

°	0	6'	12'	18'	24'	30'	36'	42'	48'	54'
	0	0.1	0.2	0.3	0.4	0.5	0.6	0.7	0.8	0.9
61	0.8746	0.8755	0.8763	0.8771	0.8780	0.8788	0.8796	0.8805	0.8813	0.8821
62	0.8829	0.8838	0.8846	0.8854	0.8862	0.8870	0.8878	0.8886	0.8894	0.8902
63	0.8910	0.8918	0.8926	0.8934	0.8942	0.8949	0.8957	0.8965	0.8973	0.8980
64	0.8988	0.8996	0.9003	0.9011	0.9018	0.9026	0.9033	0.9041	0.9048	0.9056
65	0.9063	0.9070	0.9078	0.9085	0.9092	0.9100	0.9107	0.9114	0.9121	0.9128
66	0.9135	0.9143	0.9150	0.9157	0.9164	0.9171	0.9178	0.9184	0.9191	0.9198
67	0.9205	0.9212	0.9219	0.9225	0.9232	0.9239	0.9245	0.9252	0.9259	0.9265
68	0.9272	0.9278	0.9285	0.9291	0.9298	0.9304	0.9311	0.9317	0.9323	0.9330
69	0.9336	0.9342	0.9348	0.9354	0.9361	0.9367	0.9373	0.9379	0.9385	0.9391
70	0.9397	0.9403	0.9409	0.9415	0.9421	0.9426	0.9432	0.9438	0.9444	0.9449
71	0.9455	0.9461	0.9466	0.9472	0.9478	0.9483	0.9489	0.9494	0.9500	0.9505
72	0.9511	0.9516	0.9521	0.9527	0.9532	0.9537	0.9542	0.9548	0.9553	0.9558
73	0.9563	0.9568	0.9573	0.9578	0.9583	0.9588	0.9593	0.9598	0.9603	0.9608
74	0.9613	0.9617	0.9622	0.9627	0.9632	0.9636	0.9641	0.9646	0.9650	0.9655
75	0.9659	0.9664	0.9668	0.9673	0.9677	0.9681	0.9686	0.9690	0.9694	0.9699
76	0.9703	0.9707	0.9711	0.9715	0.9720	0.9724	0.9728	0.9732	0.9736	0.9740
77	0.9744	0.9748	0.9751	0.9755	0.9759	0.9763	0.9767	0.9770	0.9774	0.9778
78	0.9781	0.9785	0.9789	0.9792	0.9796	0.9799	0.9803	0.9806	0.9810	0.9813
79	0.9816	0.9820	0.9823	0.9826	0.9829	0.9833	0.9836	0.9839	0.9842	0.9845
80	0.9848	0.9851	0.9854	0.9857	0.9860	0.9863	0.9866	0.9869	0.9871	0.9874
81	0.9877	0.9880	0.9882	0.9885	0.9888	0.9890	0.9893	0.9895	0.9898	0.9900
82	0.9903	0.9905	0.9907	0.9910	0.9912	0.9914	0.9917	0.9919	0.9921	0.9923
83	0.9925	0.9928	0.9930	0.9932	0.9934	0.9936	0.9938	0.9940	0.9942	0.9943
84	0.9945	0.9947	0.9949	0.9951	0.9952	0.9954	0.9956	0.9957	0.9959	0.9960
85	0.9962	0.9963	0.9965	0.9966	0.9968	0.9969	0.9971	0.9972	0.9973	0.9974
86	0.9976	0.9977	0.9978	0.9979	0.9980	0.9981	0.9982	0.9983	0.9984	0.9985
87	0.9986	0.9987	0.9988	0.9989	0.9990	0.9990	0.9991	0.9992	0.9993	0.9993
88	0.9994	0.9995	0.9995	0.9996	0.9996	0.9997	0.9997	0.9997	0.9998	0.9998
89	0.9998	0.9999	0.9999	0.9999	0.9999	1.0000	1.0000	1.0000	1.0000	1.0000

Table 25.2 Cosines

°	0	6'	12'	18'	24'	30'	36'	42'	48'	54'
	0	0.1	0.2	0.3	0.4	0.5	0.6	0.7	0.8	0.9
0	1.0000	1.0000	1.0000	1.0000	1.0000	1.0000	0.9999	0.9999	0.9999	0.9999
1	0.9998	0.9998	0.9998	0.9997	0.9997	0.9997	0.9996	0.9996	0.9995	0.9995
2	0.9994	0.9993	0.9993	0.9992	0.9991	0.9990	0.9990	0.9989	0.9988	0.9987
3	0.9986	0.9985	0.9984	0.9983	0.9982	0.9981	0.9980	0.9979	0.9978	0.9977
4	0.9976	0.9974	0.9073	0.9972	0.9971	0.9969	0.9968	0.9966	0.9965	0.9963
5	0.9962	0.9960	0.9959	0.9957	0.9956	0.9954	0.9952	0.9951	0.9949	0.9947
6	0.9945	0.9943	0.9942	0.9940	0.9938	0.9936	0.9934	0.9932	0.9930	0.9928
7	0.9925	0.9923	0.9921	0.9919	0.9917	0.9914	0.9912	0.9910	0.9907	0.9905
8	0.9903	0.9900	0.9898	0.9895	0.9893	0.9890	0.9888	0.9885	0.9882	0.9880
9	0.9877	0.9874	0.9871	0.9869	0.9866	0.9863	0.9860	0.9857	0.9854	0.9851
10	0.9848	0.9845	0.9842	0.9839	0.9836	0.9833	0.9829	0.9826	0.9823	0.9820
11	0.9816	0.9813	0.9810	0.9806	0.9803	0.9799	0.9796	0.9792	0.9789	0.9785
12	0.9781	0.9778	0.9774	0.9770	0.9767	0.9763	0.9759	0.9755	0.9751	0.9748
13	0.9744	0.9740	0.9736	0.9732	0.9728	0.9724	0.9720	0.9715	0.9711	0.9707
14	0.9703	0.9699	0.9694	0.9690	0.9686	0.9681	0.9677	0.9673	0.9668	0.9664
15	0.9659	0.9655	0.9650	0.9646	0.9641	0.9636	0.9632	0.9627	0.9622	0.9617
16	0.9613	0.9608	0.9603	0.9598	0.9593	0.9588	0.9583	0.9578	0.9573	0.9568
17	0.9563	0.9558	0.9553	0.9548	0.9542	0.9537	0.9532	0.9527	0.9521	0.9516
18	0.9511	0.9505	0.9500	0.9494	0.9489	0.9483	0.9478	0.9472	0.9466	0.9461
19	0.9455	0.9449	0.9444	0.9438	0.9432	0.9426	0.9421	0.9415	0.9409	0.9403
20	0.9397	0.9391	0.9385	0.9379	0.9373	0.9367	0.9361	0.9354	0.9348	0.9342
21	0.9336	0.9330	0.9323	0.9317	0.9311	0.9304	0.9298	0.9291	0.9285	0.9278
22	0.9272	0.9265	0.9259	0.9252	0.9245	0.9239	0.9232	0.9225	0.9219	0.9212
23	0.9205	0.9198	0.9191	0.9184	0.9178	0.9171	0.9164	0.9157	0.9150	0.9143
24	0.9135	0.9128	0.9121	0.9114	0.9107	0.9100	0.9092	0.9085	0.9078	0.9070
25	0.9063	0.9056	0.9048	0.9041	0.9033	0.9026	0.9018	0.9011	0.9003	0.8996
26	0.8988	0.8980	0.8973	0.8965	0.8957	0.8949	0.8942	0.8934	0.8926	0.8918
27	0.8910	0.8902	0.8894	0.8886	0.8878	0.8870	0.8862	0.8854	0.8846	0.8838
28	0.8829	0.8821	0.8813	0.8805	0.8796	0.8788	0.8780	0.8771	0.8763	0.8755
29	0.8746	0.8738	0.8729	0.8721	0.8712	0.8704	0.8695	0.8686	0.8678	0.8669
30	0.8660	0.8652	0.8643	0.8634	0.8625	0.8616	0.8607	0.8599	0.8590	0.8581
31	0.8572	0.8563	0.8554	0.8545	0.8536	0.8526	0.8517	0.8508	0.8499	0.8490

°	0	6'	12'	18'	24'	30'	36'	42'	48'	54'
	0	0.1	0.2	0.3	0.4	0.5	0.6	0.7	0.8	0.9
32	0.8480	0.8471	0.8462	0.8453	0.8443	0.8434	0.8425	0.8415	0.8406	0.8396
33	0.8387	0.8377	0.8368	0.8358	0.8348	0.8339	0.8329	0.8320	0.8310	0.8300
34	0.8290	0.8281	0.8271	0.8261	0.8251	0.8241	0.8231	0.8221	0.8211	0.8202
35	0.8192	0.8181	0.8171	0.8161	0.8151	0.8141	0.8131	0.8121	0.8111	0.8100
36	0.8090	0.8080	0.8070	0.8059	0.8049	0.8039	0.8028	0.8018	0.8007	0.7997
37	0.7986	0.7976	0.7965	0.7955	0.7944	0.7934	0.7923	0.7912	0.7902	0.7891
38	0.7880	0.7869	0.7859	0.7848	0.7837	0.7826	0.7815	0.7804	0.7793	0.7782
39	0.7771	0.7760	0.7749	0.7738	0.7727	0.7716	0.7705	0.7694	0.7683	0.7672
40	0.7660	0.7649	0.7638	0.7627	0.7615	0.7604	0.7593	0.7581	0.7570	0.7559
41	0.7547	0.7536	0.7524	0.7513	0.7501	0.7490	0.7478	0.7466	0.7455	0.7443
42	0.7431	0.7420	0.7408	0.7396	0.7385	0.7373	0.7361	0.7349	0.7337	0.7325
43	0.7314	0.7302	0.7290	0.7278	0.7266	0.7254	0.7242	0.7230	0.7218	0.7206
44	0.7193	0.7181	0.7169	0.7157	0.7145	0.7133	0.7120	0.7108	0.7096	0.7083
45	0.7071	0.7059	0.7046	0.7034	0.7022	0.7009	0.6997	0.6984	0.6972	0.6959
46	0.6947	0.6934	0.6921	0.6909	0.6896	0.6884	0.6871	0.6858	0.6845	0.6833
47	0.6820	0.6807	0.6794	0.6782	0.6769	0.6756	0.6743	0.6730	0.6717	0.6704
48	0.6691	0.6678	0.6665	0.6652	0.6639	0.6626	0.6613	0.6600	0.6587	0.6574
49	0.6561	0.6547	0.6534	0.6521	0.6508	0.6494	0.6481	0.6468	0.6455	0.6441
50	0.6428	0.6414	0.6401	0.6388	0.6374	0.6361	0.6347	0.6334	0.6320	0.6307
51	0.6293	0.6280	0.6266	0.6252	0.6239	0.6225	0.6211	0.6198	0.6184	0.6170
52	0.6157	0.6143	0.6129	0.6115	0.6101	0.6088	0.6074	0.6060	0.6046	0.6032
53	0.6018	0.6004	0.5990	0.5976	0.5962	0.5948	0.5934	0.5920	0.5906	0.5892
54	0.5878	0.5864	0.5850	0.5835	0.5821	0.5807	0.5793	0.5779	0.5764	0.5750
55	0.5736	0.5721	0.5707	0.5693	0.5678	0.5664	0.5650	0.5635	0.5621	0.5606
56	0.5592	0.5577	0.5563	0.5548	0.5534	0.5519	0.5505	0.5490	0.5476	0.5461
57	0.5446	0.5432	0.5417	0.5402	0.5388	0.5373	0.5358	0.5344	0.5329	0.5314
58	0.5299	0.5284	0.5270	0.5255	0.5240	0.5225	0.5210	0.5195	0.5180	0.5165
59	0.5150	0.5135	0.5120	0.5105	0.5090	0.5075	0.5060	0.5045	0.5030	0.5015
60	0.5000	0.4985	0.4970	0.4955	0.4939	0.4924	0.4909	0.4894	0.4879	0.4863
61	0.4848	0.4833	0.4818	0.4802	0.4787	0.4772	0.4756	0.4741	0.4726	0.4710
62	0.4695	0.4679	0.4664	0.4648	0.4633	0.4617	0.4602	0.4586	0.4571	0.4555
63	0.4540	0.4524	0.4509	0.4493	0.4478	0.4462	0.4446	0.4431	0.4415	0.4399
64	0.4384	0.4368	0.4352	0.4337	0.4321	0.4305	0.4289	0.4274	0.4258	0.4242

°	0	6'	12'	18'	24'	30'	36'	42'	48'	54'
	0	0.1	0.2	0.3	0.4	0.5	0.6	0.7	0.8	0.9
65	0.4226	0.4210	0.4195	0.4179	0.4163	0.4147	0.4131	0.4115	0.4099	0.4083
66	0.4067	0.4051	0.4035	0.4019	0.4003	0.3987	0.3971	0.3955	0.3939	0.3923
67	0.3907	0.3891	0.3875	0.3859	0.3843	0.3827	0.3811	0.3795	0.3778	0.3762
68	0.3746	0.3730	0.3714	0.3697	0.3681	0.3665	0.3649	0.3633	0.3616	0.3600
69	0.3584	0.3567	0.3551	0.3535	0.3518	0.3502	0.3486	0.3469	0.3453	0.3437
70	0.3420	0.3404	0.3387	0.3371	0.3355	0.3338	0.3322	0.3305	0.3289	0.3272
71	0.3256	0.3239	0.3223	0.3206	0.3190	0.3173	0.3156	0.3140	0.3123	0.3107
72	0.3090	0.3074	0.3057	0.3040	0.3024	0.3007	0.2990	0.2974	0.2957	0.2940
73	0.2924	0.2907	0.2890	0.2874	0.2857	0.2840	0.2823	0.2807	0.2790	0.2773
74	0.2756	0.2740	0.2723	0.2706	0.2689	0.2672	0.2656	0.2639	0.2622	0.2605
75	0.2588	0.2571	0.2554	0.2538	0.2521	0.2504	0.2487	0.2470	0.2453	0.2436
76	0.2419	0.2402	0.2385	0.2368	0.2351	0.2334	0.2317	0.2300	0.2284	0.2267
77	0.2250	0.2233	0.2215	0.2198	0.2181	0.2164	0.2147	0.2130	0.2113	0.2096
78	0.2079	0.2062	0.2045	0.2028	0.2011	0.1994	0.1977	0.1959	0.1942	0.1925
79	0.1908	0.1891	0.1874	0.1857	0.1840	0.1822	0.1805	0.1788	0.1771	0.1754
80	0.1736	0.1719	0.1702	0.1685	0.1668	0.1650	0.1633	0.1616	0.1599	0.1582
81	0.1564	0.1547	0.1530	0.1513	0.1495	0.1478	0.1461	0.1444	0.1426	0.1409
82	0.1392	0.1374	0.1357	0.1340	0.1323	0.1305	0.1288	0.1271	0.1253	0.1236
83	0.1219	0.1201	0.1184	0.1167	0.1149	0.1132	0.1115	0.1097	0.1080	0.1063
84	0.1045	0.1028	0.1011	0.0993	0.0976	0.0958	0.0941	0.0924	0.0906	0.0889
85	0.0872	0.0854	0.0837	0.0819	0.0802	0.0785	0.0767	0.0750	0.0732	0.0715
86	0.0698	0.0680	0.0663	0.0645	0.0628	0.0610	0.0593	0.0576	0.0558	0.0541
87	0.0523	0.0506	0.0488	0.0471	0.0454	0.0436	0.0419	0.0401	0.0384	0.0366
88	0.0349	0.0332	0.0314	0.0297	0.0279	0.0262	0.0244	0.0227	0.0209	0.0192
89	0.0175	0.0157	0.0140	0.0122	0.0105	0.0087	0.0070	0.0052	0.0035	0.0017

Table 25.1 3 Tangents

°	0	6'	12'	18'	24'	30'	36'	42'	48'	54'
	0	0.1	0.2	0.3	0.4	0.5	0.6	0.7	0.8	0.9
0	0	0.0017	0.0035	0.0052	0.0070	0.0087	0.0105	0.0122	0.0140	0.0157
1	0.0175	0.0192	0.0209	0.0227	0.0244	0.0262	0.0279	0.0297	0.0314	0.0332
2	0.0349	0.0367	0.0384	0.0402	0.0419	0.0437	0.0454	0.0472	0.0489	0.0507
3	0.0524	0.0542	0.0559	0.0577	0.0594	0.0612	0.0629	0.0647	0.0664	0.0682
4	0.0699	0.0717	0.0734	0.0752	0.0769	0.0787	0.0805	0.0822	0.0840	0.0857
5	0.0875	0.0892	0.0910	0.0928	0.0945	0.0963	0.0981	0.0998	0.1016	0.1033
6	0.1051	0.1069	0.1086	0.1104	0.1122	0.1139	0.1157	0.1175	0.1192	0.1210
7	0.1228	0.1246	0.1263	0.1281	0.1299	0.1317	0.1334	0.1352	0.1370	0.1388
8	0.1405	0.1423	0.1441	0.1459	0.1477	0.1495	0.1512	0.1530	0.1548	0.1566
9	0.1584	0.1602	0.1620	0.1638	0.1655	0.1673	0.1691	0.1709	0.1727	0.1745
10	0.1763	0.1781	0.1799	0.1817	0.1835	0.1853	0.1871	0.1890	0.1908	0.1926
11	0.1944	0.1962	0.1980	0.1998	0.2016	0.2035	0.2053	0.2071	0.2089	0.2107
12	0.2126	0.2144	0.2162	0.2180	0.2199	0.2217	0.2235	0.2254	0.2272	0.2290
13	0.2309	0.2327	0.2345	0.2364	0.2382	0.2401	0.2419	0.2438	0.2456	0.2475
14	0.2493	0.2512	0.2530	0.2549	0.2568	0.2586	0.2605	0.2623	0.2642	0.2661
15	0.2679	0.2698	0.2717	0.2736	0.2754	0.2773	0.2792	0.2811	0.2830	0.2849
16	0.2867	0.2886	0.2905	0.2924	0.2943	0.2962	0.2981	0.3000	0.3019	0.3038
17	0.3057	0.3076	0.3096	0.3115	0.3134	0.3153	0.3172	0.3191	0.3211	0.3230
18	0.3249	0.3269	0.3288	0.3307	0.3327	0.3346	0.3365	0.3385	0.3404	0.3424
19	0.3443	0.3463	0.3482	0.3502	0.3522	0.3541	0.3561	0.3581	0.3600	0.3620
20	0.3640	0.3659	0.3679	0.3699	0.3719	0.3739	0.3759	0.3779	0.3799	0.3819
21	0.3839	0.3859	0.3879	0.3899	0.3919	0.3939	0.3959	0.3979	0.4000	0.4020
22	0.4040	0.4061	0.4081	0.4101	0.4122	0.4142	0.4163	0.4183	0.4204	0.4224
23	0.4245	0.4265	0.4286	0.4307	0.4327	0.4348	0.4369	0.4390	0.4411	0.4431
24	0.4452	0.4473	0.4494	0.4515	0.4536	0.4557	0.4578	0.4599	0.4621	0.4642
25	0.4663	0.4684	0.4706	0.4727	0.4748	0.4770	0.4791	0.4813	0.4834	0.4856
26	0.4877	0.4899	0.4921	0.4942	0.4964	0.4986	0.5008	0.5029	0.5051	0.5073
27	0.5095	0.5117	0.5139	0.5161	0.5184	0.5206	0.5228	0.5250	0.5272	0.5295
28	0.5317	0.5340	0.5362	0.5384	0.5407	0.5430	0.5452	0.5475	0.5498	0.5520
29	0.5543	0.5566	0.5589	0.5612	0.5635	0.5658	0.5681	0.5704	0.5727	0.5750
30	0.5774	0.5797	0.5820	0.5844	0.5867	0.5890	0.5914	0.5938	0.5961	0.5985
31	0.6009	0.6032	0.6056	0.6080	0.6104	0.6128	0.6152	0.6176	0.6200	0.6224

°	0	6'	12'	18'	24'	30'	36'	42'	48'	54'
	0	0.1	0.2	0.3	0.4	0.5	0.6	0.7	0.8	0.9
32	0.6249	0.6273	0.6297	0.6322	0.6346	0.6371	0.6395	0.6420	0.6445	0.6469
33	0.6494	0.6519	0.6544	0.6569	0.6594	0.6619	0.6644	0.6669	0.6694	0.6720
34	0.6745	0.6771	0.6796	0.6822	0.6847	0.6873	0.6899	0.6924	0.6950	0.6976
35	0.7002	0.7028	0.7054	0.7080	0.7107	0.7133	0.7159	0.7186	0.7212	0.7239
36	0.7265	0.7292	0.7319	0.7346	0.7373	0.7400	0.7427	0.7454	0.7481	0.7508
37	0.7536	0.7563	0.7590	0.7618	0.7646	0.7673	0.7701	0.7729	0.7757	0.7785
38	0.7813	0.7841	0.7869	0.7898	0.7926	0.7954	0.7983	0.8012	0.8040	0.8069
39	0.8098	0.8127	0.8156	0.8185	0.8214	0.8243	0.8273	0.8302	0.8332	0.8361
40	0.8391	0.8421	0.8451	0.8481	0.8511	0.8541	0.8571	0.8601	0.8632	0.8662
41	0.8693	0.8724	0.8754	0.8785	0.8816	0.8847	0.8878	0.8910	0.8941	0.8972
42	0.9004	0.9036	0.9067	0.9099	0.9131	0.9163	0.9195	0.9228	0.9260	0.9293
43	0.9325	0.9358	0.9391	0.9424	0.9457	0.9490	0.9523	0.9556	0.9590	0.9623
44	0.9657	0.9691	0.9725	0.9759	0.9793	0.9827	0.9861	0.9896	0.9930	0.9965
45	1.0000	1.0035	1.0070	1.0105	1.0141	1.0176	1.0212	1.0247	1.0283	1.0319
46	1.0355	1.0392	1.0428	1.0464	1.0501	1.0538	1.0575	1.0612	1.0649	1.0686
47	1.0724	1.0761	1.0799	1.0837	1.0875	1.0913	1.0951	1.0990	1.1028	1.1067
48	1.1106	1.1145	1.1184	1.1224	1.1263	1.1303	1.1343	1.1383	1.1423	1.1463
49	1.1504	1.1544	1.1585	1.1626	1.1667	1.1708	1.1750	1.1792	1.1833	1.1875
50	1.1918	1.1960	1.2002	1.2045	1.2088	1.2131	1.2174	1.2218	1.2261	1.2305
51	1.2349	1.2393	1.2437	1.2482	1.2527	1.2572	1.2617	1.2662	1.2708	1.2753
52	1.2799	1.2846	1.2892	1.2938	1.2985	1.3032	1.3079	1.3127	1.3175	1.3222
53	1.3270	1.3319	1.3367	1.3416	1.3465	1.3514	1.3564	1.3613	1.3663	1.3713
54	1.3764	1.3814	1.3865	1.3916	1.3968	1.4019	1.4071	1.4124	1.4176	1.4229
55	1.4281	1.4335	1.4388	1.4442	1.4496	1.4550	1.4605	1.4659	1.4715	1.4770
56	1.4826	1.4882	1.4938	1.4994	1.5051	1.5108	1.5166	1.5224	1.5282	1.5340
57	1.5399	1.5458	1.5517	1.5577	1.5637	1.5697	1.5757	1.5818	1.5880	1.5941
58	1.6003	1.6066	1.6128	1.6191	1.6255	1.6319	1.6383	1.6447	1.6512	1.6577
59	1.6643	1.6709	1.6775	1.6842	1.6909	1.6977	1.7045	1.7113	1.7182	1.7251
60	1.7321	1.7391	1.7461	1.7532	1.7603	1.7675	1.7747	1.7820	1.7893	1.7966
61	1.8040	1.8115	1.8190	1.8265	1.8341	1.8418	1.8495	1.8572	1.8650	1.8728
62	1.8807	1.8887	1.8967	1.9047	1.9128	1.9210	1.9292	1.9375	1.9458	1.9542
63	1.9626	1.9711	1.9797	1.9883	1.9970	2.0057	2.0145	2.0233	2.0323	2.0413
64	2.0503	2.0594	2.0686	2.0778	2.0872	2.0965	2.1060	2.1155	2.1251	2.1348
65	2.1445	2.1543	2.1642	2.1742	2.1842	2.1943	2.2045	2.2148	2.2251	2.2355

°	0	6'	12'	18'	24'	30'	36'	42'	48'	54'
	0	0.1	0.2	0.3	0.4	0.5	0.6	0.7	0.8	0.9
66	2.2460	2.2566	2.2673	2.2781	2.2889	2.2998	2.3109	2.3220	2.3332	2.3445
67	2.3559	2.3673	2.3789	2.3906	2.4023	2.4142	2.4262	2.4383	2.4504	2.4627
68	2.4751	2.4876	2.5002	2.5129	2.5257	2.5386	2.5517	2.5649	2.5782	2.5916
69	2.6051	2.6187	2.6325	2.6464	2.6605	2.6746	2.6889	2.7034	2.7179	2.7326
70	2.7475	2.7625	2.7776	2.7929	2.8083	2.8239	2.8397	2.8556	2.8716	2.8878
71	2.9042	2.9208	2.9375	2.9544	2.9714	2.9887	3.0061	3.0237	3.0415	3.0595
72	3.0777	3.0961	3.1146	3.1334	3.1524	3.1716	3.1910	3.2106	3.2305	3.2506
73	3.2709	3.2914	3.3122	3.3332	3.3544	3.3759	3.3977	3.4197	3.4420	3.4646
74	3.4874	3.5105	3.5339	3.5576	3.5816	3.6059	3.6305	3.6554	3.6806	3.7062
75	3.7321	3.7583	3.7848	3.8118	3.8391	3.8667	3.8947	3.9232	3.9520	3.9812
76	4.0108	4.0408	4.0713	4.1022	4.1335	4.1653	4.1976	4.2303	4.2635	4.2972
77	4.3315	4.3662	4.4015	4.4373	4.4737	4.5107	4.5483	4.5864	4.6252	4.6646
78	4.7046	4.7453	4.7867	4.8288	4.8716	4.9152	4.9594	5.0045	5.0504	5.0970
79	5.1446	5.1929	5.2422	5.2924	5.3435	5.3955	5.4486	5.5026	5.5578	5.6140
80	5.671	5.730	5.789	5.850	5.912	5.976	6.041	6.107	6.174	6.243
81	6.314	6.386	6.460	6.535	6.612	6.691	6.772	6.855	6.940	7.026
82	7.115	7.207	7.300	7.396	7.495	7.596	7.700	7.806	7.916	8.028
83	8.144	8.264	8.386	8.513	8.643	8.777	8.915	9.058	9.205	9.357
84	9.514	9.677	9.845	10.019	10.199	10.385	10.579	10.780	10.988	11.205
85	11.430	11.664	11.909	12.163	12.429	12.706	12.996	13.300	13.617	13.951
86	14.301	14.669	15.056	15.464	15.895	16.350	16.832	17.343	17.886	18.464
87	19.081	19.740	20.446	21.205	22.022	22.904	23.859	24.898	26.031	27.271
88	28.636	30.145	31.821	33.694	35.801	38.188	40.917	44.066	47.739	52.081
89	57.29	63.66	71.62	81.85	95.49	114.59	143.24	190.98	286.48	572.96

CHAPTER 26
PRIME NUMBERS

PRIME NUMBERS AND FACTORS OF NON-PRIME NUMBERS

PRIME NUMBERS

These are numbers that are only divisible by 1 and by the number itself, e.g.

$$13 = 1 \times 13$$
$$29 = 1 \times 29$$

NON-PRIME NUMBERS

These are numbers which can be divided by two or more numbers, in addition to 1 and the number itself.

PRIME FACTORS

These are prime numbers that, when multiplied together, give the number being considered, e.g. the prime factors of 165 are 3, 5, and 11, i.e.:

$$165 = 3 \times 5 \times 11$$

Prime factors of odd numbers

The tables give the lowest prime factor for odd numbers, but by the following process all prime factors can be found.

For example, to find the prime factors of 165:

1. Look up 165 and the smallest factor is 3.

2. Divide 165 by 3 to give 55.

3. Look up 55 and the smallest factor is 5.

4. Divide 55 by 5 to give 11.

5. Look up 11, which is found to be a prime number.

6. The prime factors are therefore 3, 5 and 11, i.e.:

$$165 = 3 \times 5 \times 11$$

Prime factors of even numbers

These can be found as follows. The smallest prime factor for any even number is 2. Therefore, divide the number by 2 and, if this gives an odd number, use the tables for the remaining prime number factors, otherwise divide by 2 again.

For example, to find the prime factors of 84:

1. Being an even number the smallest factor is 2.

2. Divide 84 by 2 to give 42.

3. Being an even number the smallest factor is 2.

4. Divide 42 by 2 to give 21.

5. 21 is an odd number.

6. Look up 21 and the smallest factor is 3.

7. Divide 21 by 3 to give 7.

8. Look up 7, which is found to be a prime number.

9. The prime factors are therefore 2, 2, 3 and 7, i.e.

$$84 = 2 \times 2 \times 3 \times 7$$

Table 26.1 Prime numbers and smallest factors: part 1

	1	101	201	301	401	501	601	701	801	901
	up	up	up	up	up	up	up	up	up	up
1	P	P	3	7	P	3	P	P	3	17
3	P	P	7	3	13	P	3	19	11	3
5	P	3	5	5	3	5	5	3	5	5
7	P	P	3	P	11	3	P	7	3	P
9	3	P	11	3	P	P	3	P	P	3
11	P	3	P	P	3	7	13	3	P	P
13	P	P	3	P	7	3	P	23	3	11
15	3	5	5	3	5	5	3	5	5	3
17	P	3	7	P	3	11	P	3	19	7
19	P	7	3	11	P	3	P	P	3	P
21	3	11	13	3	P	P	3	7	P	3
23	P	3	P	17	3	P	7	3	P	13
25	5	5	3	5	5	3	5	5	3	5
27	3	P	P	3	7	17	3	P	P	3
29	P	3	P	7	3	23	17	3	P	P
31	P	P	3	P	P	3	P	17	3	7
33	3	7	P	3	P	13	3	P	7	3
35	5	3	5	5	3	5	5	3	5	5
37	P	P	3	P	19	3	7	11	3	P
39	3	P	P	3	P	7	3	P	P	3
41	P	3	P	11	3	P	P	3	29	P
43	P	11	3	7	P	3	P	P	3	23
45	3	5	5	3	5	5	3	5	5	3
47	P	3	13	P	3	P	P	3	7	P
49	7	P	3	P	P	3	11	7	3	13

Table 26.2 Prime numbers and smallest factors: part 2

	51	151	251	351	451	551	651	751	851	951
	up	up	up	up	up	up	up	up	up	up
51	3	P	P	3	11	19	3	P	23	3
53	P	3	11	P	3	7	P	3	P	P
55	5	5	3	5	5	3	5	5	3	5
57	3	P	P	3	P	P	3	P	P	3
59	P	3	7	P	3	13	P	3	P	7
61	P	7	3	19	P	3	P	P	3	31
63	3	P	P	3	P	P	3	7	P	3
65	5	3	5	5	3	5	5	3	5	5
67	P	P	3	P	P	3	23	13	3	P
69	3	13	P	3	7	P	3	P	11	3

	51	151	251	351	451	551	651	751	851	951
	up	up	up	up	up	up	up	up	up	up
71	P	3	P	7	3	P	11	3	13	P
73	P	P	3	P	11	3	P	P	3	7
75	3	5	5	3	5	5	3	5	5	3
77	7	3	P	13	3	P	P	3	P	P
79	P	P	3	P	P	3	7	19	3	11
81	3	P	P	3	13	7	3	11	P	3
83	P	3	P	P	3	11	P	3	P	P
85	5	5	3	5	5	3	5	5	3	5
87	3	11	7	3	P	P	3	P	P	3
89	P	3	17	P	3	19	13	3	7	23
91	7	P	3	17	P	3	P	7	3	P
93	3	P	P	3	17	P	3	13	19	3
95	5	3	5	5	3	5	5	3	5	5
97	P	P	3	P	7	3	17	P	3	P
99	3	P	13	3	P	P	3	17	29	3

Table 26.3 Prime numbers and smallest factors: part 3

	1001	1101	1201	1301	1401	1501	1601	1701	1801	1901
	up	up	up	up	up	up	up	up	up	up
1	7	3	P	P	3	19	P	3	P	P
3	17	P	3	P	23	3	7	13	3	11
5	3	5	5	3	5	5	3	5	5	3
7	19	3	17	P	3	11	P	3	13	P
9	P	P	3	7	P	3	P	P	3	23
11	3	11	7	3	17	P	3	29	P	3
13	P	3	P	13	3	17	P	3	7	P
15	5	5	3	5	5	3	5	5	3	5
17	3	P	P	3	13	37	3	17	23	3
19	P	3	23	P	3	7	P	3	17	19
21	P	19	3	P	7	3	P	P	3	17
23	3	P	P	3	P	P	3	P	P	3
25	5	3	5	5	3	5	5	3	5	5
27	13	7	3	P	P	3	P	11	3	41
29	3	P	P	3	P	11	3	7	31	3
31	P	3	P	11	3	P	7	3	P	P
33	P	11	3	31	P	3	23	P	3	P
35	3	5	5	3	5	5	3	5	5	3
37	17	3	P	7	3	29	P	3	11	13
39	P	17	3	13	P	3	11	37	3	7

	1001	1101	1201	1301	1401	1501	1601	1701	1801	1901
	up	up	up	up	up	up	up	up	up	up
41	3	7	17	3	11	23	3	P	7	3
43	7	3	11	17	3	P	31	3	19	29
45	5	5	3	5	5	3	5	5	3	5
47	3	31	29	3	P	7	3	P	P	3
49	P	3	P	19	3	P	17	3	43	P

Table 26.4 Prime numbers and smallest factors: part 4

	1051	1151	1251	1351	1451	1515	1651	1751	1851	1951
	up	up	up	up	up	up	up	up	up	up
51	P	P	3	7	P	3	13	17	3	P
53	3	P	7	3	P	P	3	P	17	3
55	5	3	5	5	3	5	5	3	5	5
57	7	13	3	23	31	3	P	7	3	19
59	3	19	P	3	P	P	3	P	11	3
61	P	3	13	P	3	7	11	3	P	37
63	P	P	3	29	7	3	P	41	3	13
65	3	5	5	3	5	5	3	5	5	3
67	11	3	7	P	3	P	P	3	P	7
69	P	7	3	37	13	3	P	29	3	11
71	3	P	31	3	P	P	3	7	P	3
73	29	3	19	P	3	11	7	3	P	P
75	5	5	3	5	5	3	5	5	3	5
77	3	11	P	3	7	19	3	P	P	3
79	13	3	P	7	3	P	23	3	P	P
81	23	P	3	P	P	3	41	13	3	7
83	3	7	P	3	P	P	3	P	7	3
85	5	3	5	5	3	5	5	3	5	5
87	P	P	3	19	P	3	7	P	3	P
89	3	29	P	3	P	7	3	P	P	3
91	P	3	P	13	3	37	19	3	31	11
93	P	P	3	7	P	3	P	11	3	P
95	3	5	5	3	5	5	3	5	5	3
97	P	3	P	11	3	P	P	3	7	P
99	7	11	3	P	P	3	P	7	3	P

Table 26.5 Prime numbers from 2000 to 5999

2003	11	17	27	29	39	53	63	69	81	83	87	89	99	-	-
2111	13	29	31	37	41	43	53	61	79	-	-	-	-	-	-
2203	07	13	21	37	39	43	51	67	69	73	81	87	93	97	-
2309	11	33	39	41	47	51	57	71	77	81	83	89	93	99	-
2411	17	23	37	41	47	59	67	73	77	-	-	-	-	-	-
2503	21	31	39	43	49	51	57	79	91	93	-	-	-	-	-
2609	17	21	33	47	57	59	63	71	77	83	87	89	93	99	-
2707	11	13	19	29	31	41	49	53	67	77	89	91	97	-	-
2801	03	19	33	37	43	51	57	61	79	87	97	-	-	-	-
2903	09	17	27	39	53	57	63	69	71	99	-	-	-	-	-
3001	11	19	23	37	41	49	61	67	79	83	89	-	-	-	-
3109	19	21	37	63	67	69	81	87	91	-	-	-	-	-	-
3203	09	17	21	29	51	53	57	59	71	99	-	-	-	-	-
3301	07	13	19	23	29	31	43	47	59	61	71	73	89	91	-
3407	13	33	49	57	61	63	67	69	91	99	-	-	-	-	-
3511	17	27	29	33	39	41	47	57	59	71	81	83	93	-	-
3607	13	17	23	31	37	43	59	71	73	77	91	97	-	-	-
3701	09	19	27	33	39	61	67	69	79	93	97	-	-	-	-
3803	21	23	33	47	51	53	63	77	81	89	-	-	-	-	-
3907	11	17	19	23	29	31	43	47	67	89	-	-	-	-	-
4001	03	07	13	19	21	27	49	51	57	73	79	91	93	99	-
4111	27	29	33	39	53	57	59	77	-	-	-	-	-	-	-
4201	11	17	19	29	31	41	43	53	59	61	71	73	83	89	97
4327	37	39	49	57	63	73	91	97	-	-	-	-	-	-	-
4409	21	23	41	47	51	57	63	81	83	93	-	-	-	-	-
4507	13	17	19	23	47	49	61	67	83	91	97	-	-	-	-
4603	21	37	39	43	49	51	57	63	73	79	91	-	-	-	-
4703	21	23	29	33	51	59	83	87	89	93	99	-	-	-	-
4801	13	17	31	61	71	77	89	-	-	-	-	-	-	-	-
4903	09	19	31	33	37	43	51	57	67	69	73	87	93	99	-

Table 26.6 Prime numbers from 5000 to 7999

5003	09	11	21	23	39	51	59	77	81	87	99	-	-	-	-
5101	07	13	19	47	53	67	71	79	89	97	-	-	-	-	-
5209	27	31	33	37	61	73	79	81	97	-	-	-	-	-	-
5303	09	23	33	47	51	81	87	93	99	-	-	-	-	-	-
5407	13	17	19	31	37	41	43	49	71	77	79	83	-	-	-
5501	03	07	19	21	27	31	57	63	69	73	81	91	-	-	-
5623	39	41	47	51	53	57	59	69	83	89	93	-	-	-	-
5701	11	17	37	41	43	49	79	83	91	-	-	-	-	-	-
5801	07	13	21	27	39	43	49	51	57	61	67	69	79	81	97
5903	23	27	39	53	81	87	-	-	-	-	-	-	-	-	-
6007	11	29	37	43	47	53	67	73	79	89	91	-	-	-	-
6101	13	21	31	33	43	51	63	73	97	99	-	-	-	-	-
6203	11	17	21	29	47	57	63	69	71	77	87	99	-	-	-
6301	11	17	23	29	37	43	53	59	61	67	73	79	89	97	-
6421	27	49	51	69	73	81	91	-	-	-	-	-	-	-	-
6521	29	47	51	53	63	69	71	77	81	99	-	-	-	-	-
6607	19	37	53	59	61	73	79	89	91	-	-	-	-	-	-
6701	03	09	19	33	37	61	63	79	81	91	93	-	-	-	-
6803	23	27	29	33	41	57	63	69	71	83	99	-	-	-	-
6907	11	17	47	49	59	61	67	71	77	83	91	97	-	-	-
7001	13	19	27	39	43	57	69	79	-	-	-	-	-	-	-
7103	09	21	27	29	51	59	77	87	93	-	-	-	-	-	-
7207	11	13	19	29	37	43	47	53	83	97	-	-	-	-	-
7307	09	21	31	33	49	51	69	93	-	-	-	-	-	-	-
7411	17	33	51	57	59	77	81	87	89	99	-	-	-	-	-
7507	17	23	29	37	41	47	49	59	61	73	77	83	89	91	-
7603	07	21	39	43	49	69	73	81	87	91	99	-	-	-	-
7703	17	23	27	41	53	57	59	89	93	-	-	-	-	-	-
7817	23	29	41	53	67	73	77	79	83	-	-	-	-	-	-
7901	07	19	27	33	37	49	51	63	93	-	-	-	-	-	-

CHAPTER 26: PRIME NUMBERS

Table 26.7 Prime numbers from 8000 to 9999

8009	11	17	39	53	59	69	81	87	89	93	-	-	-	-
8101	11	17	23	47	61	67	71	79	91	-	-	-	-	-
8209	19	21	31	33	37	43	63	69	73	87	91	93	97	
8311	17	29	53	63	69	77	87	89	-	-	-	-	-	-
8419	23	29	31	43	47	61	67	-	-	-	-	-	-	-
8501	13	21	27	37	39	43	63	73	81	97	99	-	-	-
8609	23	27	29	41	47	63	69	77	81	89	93	99	-	-
8707	13	19	31	37	41	47	53	61	79	83	-	-	-	-
8803	07	19	21	31	37	39	49	61	63	67	87	93	-	-
8923	29	33	41	51	63	69	71	99	-	-	-	-	-	-
9001	07	11	13	29	41	43	49	59	67	91	-	-	-	-
9103	09	27	33	37	51	57	61	73	81	87	99	-	-	-
9203	09	21	27	39	41	57	77	81	83	93	-	-	-	-
9311	19	23	37	41	43	49	71	77	91	97	-	-	-	-
9403	13	19	21	31	33	37	39	61	63	67	73	79	91	97
9511	21	33	39	47	51	87	-	-	-	-	-	-	-	-
9601	13	19	23	29	31	43	49	61	77	79	89	97	-	-
9719	21	33	39	43	49	67	69	81	87	91	-	-	-	-
9803	11	17	29	33	39	51	57	59	71	83	87	-	-	-
9901	07	23	29	31	41	49	67	73	-	-	-	-	-	-

NUMBERS GREATER THAN 1999

Having chosen to publish prime numbers and prime factors for non-prime numbers for four-digit numbers, ideally the list should go up to 9999, but this would take more space than can be justified. Therefore the second part of the listing only includes details of prime numbers between 2000 and 9999. The following procedure will though enable factors for non prime numbers between 2000 and 9999 to be found with relative ease.

Using a calculator, and for odd numbers, first attempt to divide the number in question by 3. If the result is a whole number, 3 is the lowest prime factor. If the result is not a whole number, try dividing it by 5 followed by 7, and then by the remaining prime numbers through to 97 until a whole number results.

Once you have found the lowest prime factor, divide the number by that factor. In most cases the result will be less than 1999 and the tables can then be used to find further factors. If not, repeat the above process after checking whether the number is a prime number.

The procedure will rarely be as time-consuming as may first be thought because every third number is divisible by 3, every fifth number by 5, every seventh number by 7 and so on; as a result most numbers have low prime factors. Note that no number in this range has a lowest prime factor greater than 97.

USING PRIME FACTORS

The most probable use for prime factor tables is in the design of gear chains requiring complex ratios, such as those required when determining changewheel combinations for cutting metric threads on a lathe with an imperial leadscrew, and vice versa. They are also used when cutting wormwheels, which also require complex ratios.

EXAMPLE

If you wish to cut a thread with a pitch of 1mm on a lathe with an 8 TPI lead screw, the lathe mandrel will have to rotate 25.4 times while the leadscrew rotates 8 times. This therefore requires a ratio of 25.4 : 8. Expressing this as a fraction we get:

$$\frac{25.4}{8} = \frac{1}{0.314960629}$$

Accepting that an exact ratio is not possible, this can be simplified to

$$\frac{1}{0.315}$$

As decimals are not appropriate to fractions this must be written as:

$$\frac{1000}{315}$$

Factorizing both 1000 and 315, using the tables, we get:

$$\frac{2 \times 2 \times 2 \times 5 \times 5 \times 5}{3 \times 3 \times 5 \times 7}$$

Cancelling out one 5 we get:

$$\frac{2 \times 2 \times 2 \times 5 \times 5}{3 \times 3 \times 7}$$

Ignoring the fact that gears cannot be made with so few teeth, we still have a problem as there are 5 driven gears but only 3 drivers. If we multiply 2 x 2 and 2 x 5 we get 3 drivers and 3 driven as follows:

$$\frac{4 \times 10 \times 5}{3 \times 3 \times 7}$$

Multiplying each number by 10 will give practical gear sizes and, being in tens, these are likely to be available with the lathe. We get:

$$\frac{40 \times 100 \times 50}{30 \times 30 \times 70}$$

A 100-tooth gear is unlikely to be available so reducing this to 50, and the 70 to 35, will retain the ratio:

$$\frac{40 \times 50 \times 50}{30 \times 30 \times 35}$$

The combination now requires two 30-tooth and two 50-tooth gears, but increasing one of each by a factor of 1.5 they become 45- and 75-tooth gears.

The resulting gear chain becomes:

$$\frac{40 \times 50 \times 75}{30 \times 35 \times 45}$$

This gives a TPI as follows:

$$8 \times \frac{40 \times 50 \times 75}{30 \times 35 \times 45} = 25.3968254$$

Pitch in millimeters is 1.000125mm, giving an error of only 0.000125mm:

$$\frac{25.4}{25.3968254} = 1.000125\text{mm}$$

DESIGNING A GEARBOX

If we need to design a gearbox with a given ratio from scratch, and using gears made specially for the application, the procedure is somewhat easier because we are no longer restricted by the limited range of gear sizes that would be available with the lathe.

In order to compare the methods, we will attempt to produce a gearbox with the same ratio as the above.

Starting with:

$$\frac{1000}{315} \quad \text{(as above)}$$

By factorizing both 1000 and 315, using the tables, we get:

$$\frac{2 \times 2 \times 2 \times 5 \times 5 \times 5}{3 \times 3 \times 5 \times 7}$$

Now, assuming a gear chain of 2 drivers and 2 driven, we multiply the numbers together using two groups above and below the line. The following is typical, but any grouping can be tried:

$$\frac{(2 \times 2 \times 5) \times (2 \times 5 \times 5)}{(3 \times 5) \times (3 \times 7)}$$

which gives gear sizes of:

$$\frac{20 \times 50}{15 \times 21}$$

Gear sizes can of course be changed, provided that the top to bottom ratio is maintained, Typically, if you prefer not to have gears as small as 15 teeth then this can be doubled in size provided that one of the gears on the top line is also doubled in size, thus retaining the ratio.

If you can adopt a single driver single driven set-up then one 5 can be cancelled out from the top and bottom lines, giving gear sizes of:

$$\frac{2 \times 2 \times 2 \times 5 \times 5}{3 \times 3 \times 7} = \frac{200}{63}$$

CHAPTER 27

CONVERSION TABLES AND FACTORS

CONVERSION TABLES

Table 27.1. Millimeters to inches: 0 to 25.4mm in 0.01mm stages

mm	0.00	0.01	0.02	0.03	0.04	0.05	0.06	0.07	0.08	0.09
	Inches									
0	0.0000	0.0004	0.0008	0.0012	0.0016	0.0020	0.0024	0.0028	0.0031	0.0035
0.10	0.0039	0.0043	0.0047	0.0051	0.0055	0.0059	0.0063	0.0067	0.0071	0.0075
0.20	0.0079	0.0083	0.0087	0.0091	0.0094	0.0098	0.0102	0.0106	0.0110	0.0114
0.30	0.0118	0.0122	0.0126	0.0130	0.0134	0.0138	0.0142	0.0146	0.0150	0.0154
0.40	0.0157	0.0161	0.0165	0.0169	0.0173	0.0177	0.0181	0.0185	0.0189	0.0193
0.50	0.0197	0.0201	0.0205	0.0209	0.0213	0.0217	0.0220	0.0224	0.0228	0.0232
0.60	0.0236	0.0240	0.0244	0.0248	0.0252	0.0256	0.0260	0.0264	0.0268	0.0272
0.70	0.0276	0.0280	0.0283	0.0287	0.0291	0.0295	0.0299	0.0303	0.0307	0.0311
0.80	0.0315	0.0319	0.0323	0.0327	0.0331	0.0335	0.0339	0.0343	0.0346	0.0350
0.90	0.0354	0.0358	0.0362	0.0366	0.0370	0.0374	0.0378	0.0382	0.0386	0.0390
1.00	0.0394	0.0398	0.0402	0.0406	0.0409	0.0413	0.0417	0.0421	0.0425	0.0429
1.10	0.0433	0.0437	0.0441	0.0445	0.0449	0.0453	0.0457	0.0461	0.0465	0.0469
1.20	0.0472	0.0476	0.0480	0.0484	0.0488	0.0492	0.0496	0.0500	0.0504	0.0508
1.30	0.0512	0.0516	0.0520	0.0524	0.0528	0.0531	0.0535	0.0539	0.0543	0.0547
1.40	0.0551	0.0555	0.0559	0.0563	0.0567	0.0571	0.0575	0.0579	0.0583	0.0587
1.50	0.0591	0.0594	0.0598	0.0602	0.0606	0.0610	0.0614	0.0618	0.0622	0.0626
1.60	0.0630	0.0634	0.0638	0.0642	0.0646	0.0650	0.0654	0.0657	0.0661	0.0665
1.70	0.0669	0.0673	0.0677	0.0681	0.0685	0.0689	0.0693	0.0697	0.0701	0.0705
1.80	0.0709	0.0713	0.0717	0.0720	0.0724	0.0728	0.0732	0.0736	0.0740	0.0744
1.90	0.0748	0.0752	0.0756	0.0760	0.0764	0.0768	0.0772	0.0776	0.0780	0.0783
2.00	0.0787	0.0791	0.0795	0.0799	0.0803	0.0807	0.0811	0.0815	0.0819	0.0823
2.10	0.0827	0.0831	0.0835	0.0839	0.0843	0.0846	0.0850	0.0854	0.0858	0.0862
2.20	0.0866	0.0870	0.0874	0.0878	0.0882	0.0886	0.0890	0.0894	0.0898	0.0902
2.30	0.0906	0.0909	0.0913	0.0917	0.0921	0.0925	0.0929	0.0933	0.0937	0.0941
2.40	0.0945	0.0949	0.0953	0.0957	0.0961	0.0965	0.0969	0.0972	0.0976	0.0980
2.50	0.0984	0.0988	0.0992	0.0996	0.1000	0.1004	0.1008	0.1012	0.1016	0.1020
2.60	0.1024	0.1028	0.1031	0.1035	0.1039	0.1043	0.1047	0.1051	0.1055	0.1059

mm	0.00	0.01	0.02	0.03	0.04	0.05	0.06	0.07	0.08	0.09
	Inches									
2.70	0.1063	0.1067	0.1071	0.1075	0.1079	0.1083	0.1087	0.1091	0.1094	0.1098
2.80	0.1102	0.1106	0.1110	0.1114	0.1118	0.1122	0.1126	0.1130	0.1134	0.1138
2.90	0.1142	0.1146	0.1150	0.1154	0.1157	0.1161	0.1165	0.1169	0.1173	0.1177
3.00	0.1181	0.1185	0.1189	0.1193	0.1197	0.1201	0.1205	0.1209	0.1213	0.1217
3.10	0.1220	0.1224	0.1228	0.1232	0.1236	0.1240	0.1244	0.1248	0.1252	0.1256
3.20	0.1260	0.1264	0.1268	0.1272	0.1276	0.1280	0.1283	0.1287	0.1291	0.1295
3.30	0.1299	0.1303	0.1307	0.1311	0.1315	0.1319	0.1323	0.1327	0.1331	0.1335
3.40	0.1339	0.1343	0.1346	0.1350	0.1354	0.1358	0.1362	0.1366	0.1370	0.1374
3.50	0.1378	0.1382	0.1386	0.1390	0.1394	0.1398	0.1402	0.1406	0.1409	0.1413
3.60	0.1417	0.1421	0.1425	0.1429	0.1433	0.1437	0.1441	0.1445	0.1449	0.1453
3.70	0.1457	0.1461	0.1465	0.1469	0.1472	0.1476	0.1480	0.1484	0.1488	0.1492
3.80	0.1496	0.1500	0.1504	0.1508	0.1512	0.1516	0.1520	0.1524	0.1528	0.1531
3.90	0.1535	0.1539	0.1543	0.1547	0.1551	0.1555	0.1559	0.1563	0.1567	0.1571
4.00	0.1575	0.1579	0.1583	0.1587	0.1591	0.1594	0.1598	0.1602	0.1606	0.1610
4.10	0.1614	0.1618	0.1622	0.1626	0.1630	0.1634	0.1638	0.1642	0.1646	0.1650
4.20	0.1654	0.1657	0.1661	0.1665	0.1669	0.1673	0.1677	0.1681	0.1685	0.1689
4.30	0.1693	0.1697	0.1701	0.1705	0.1709	0.1713	0.1717	0.1720	0.1724	0.1728
4.40	0.1732	0.1736	0.1740	0.1744	0.1748	0.1752	0.1756	0.1760	0.1764	0.1768
4.50	0.1772	0.1776	0.1780	0.1783	0.1787	0.1791	0.1795	0.1799	0.1803	0.1807
4.60	0.1811	0.1815	0.1819	0.1823	0.1827	0.1831	0.1835	0.1839	0.1843	0.1846
4.70	0.1850	0.1854	0.1858	0.1862	0.1866	0.1870	0.1874	0.1878	0.1882	0.1886
4.80	0.1890	0.1894	0.1898	0.1902	0.1906	0.1909	0.1913	0.1917	0.1921	0.1925
4.90	0.1929	0.1933	0.1937	0.1941	0.1945	0.1949	0.1953	0.1957	0.1961	0.1965
5.00	0.1969	0.1972	0.1976	0.1980	0.1984	0.1988	0.1992	0.1996	0.2000	0.2004
5.10	0.2008	0.2012	0.2016	0.2020	0.2024	0.2028	0.2031	0.2035	0.2039	0.2043
5.20	0.2047	0.2051	0.2055	0.2059	0.2063	0.2067	0.2071	0.2075	0.2079	0.2083
5.30	0.2087	0.2091	0.2094	0.2098	0.2102	0.2106	0.2110	0.2114	0.2118	0.2122
5.40	0.2126	0.2130	0.2134	0.2138	0.2142	0.2146	0.2150	0.2154	0.2157	0.2161
5.50	0.2165	0.2169	0.2173	0.2177	0.2181	0.2185	0.2189	0.2193	0.2197	0.2201
5.60	0.2205	0.2209	0.2213	0.2217	0.2220	0.2224	0.2228	0.2232	0.2236	0.2240
5.70	0.2244	0.2248	0.2252	0.2256	0.2260	0.2264	0.2268	0.2272	0.2276	0.2280
5.80	0.2283	0.2287	0.2291	0.2295	0.2299	0.2303	0.2307	0.2311	0.2315	0.2319
5.90	0.2323	0.2327	0.2331	0.2335	0.2339	0.2343	0.2346	0.2350	0.2354	0.2358
6.00	0.2362	0.2366	0.2370	0.2374	0.2378	0.2382	0.2386	0.2390	0.2394	0.2398
6.10	0.2402	0.2406	0.2409	0.2413	0.2417	0.2421	0.2425	0.2429	0.2433	0.2437
6.20	0.2441	0.2445	0.2449	0.2453	0.2457	0.2461	0.2465	0.2469	0.2472	0.2476
6.30	0.2480	0.2484	0.2488	0.2492	0.2496	0.2500	0.2504	0.2508	0.2512	0.2516
6.40	0.2520	0.2524	0.2528	0.2531	0.2535	0.2539	0.2543	0.2547	0.2551	0.2555

mm	0.00	0.01	0.02	0.03	0.04	0.05	0.06	0.07	0.08	0.09
					Inches					
6.50	0.2559	0.2563	0.2567	0.2571	0.2575	0.2579	0.2583	0.2587	0.2591	0.2594
6.60	0.2598	0.2602	0.2606	0.2610	0.2614	0.2618	0.2622	0.2626	0.2630	0.2634
6.70	0.2638	0.2642	0.2646	0.2650	0.2654	0.2657	0.2661	0.2665	0.2669	0.2673
6.80	0.2677	0.2681	0.2685	0.2689	0.2693	0.2697	0.2701	0.2705	0.2709	0.2713
6.90	0.2717	0.2720	0.2724	0.2728	0.2732	0.2736	0.2740	0.2744	0.2748	0.2752
7.00	0.2756	0.2760	0.2764	0.2768	0.2772	0.2776	0.2780	0.2783	0.2787	0.2791
7.10	0.2795	0.2799	0.2803	0.2807	0.2811	0.2815	0.2819	0.2823	0.2827	0.2831
7.20	0.2835	0.2839	0.2843	0.2846	0.2850	0.2854	0.2858	0.2862	0.2866	0.2870
7.30	0.2874	0.2878	0.2882	0.2886	0.2890	0.2894	0.2898	0.2902	0.2906	0.2909
7.40	0.2913	0.2917	0.2921	0.2925	0.2929	0.2933	0.2937	0.2941	0.2945	0.2949
7.50	0.2953	0.2957	0.2961	0.2965	0.2969	0.2972	0.2976	0.2980	0.2984	0.2988
7.60	0.2992	0.2996	0.3000	0.3004	0.3008	0.3012	0.3016	0.3020	0.3024	0.3028
7.70	0.3031	0.3035	0.3039	0.3043	0.3047	0.3051	0.3055	0.3059	0.3063	0.3067
7.80	0.3071	0.3075	0.3079	0.3083	0.3087	0.3091	0.3094	0.3098	0.3102	0.3106
7.90	0.3110	0.3114	0.3118	0.3122	0.3126	0.3130	0.3134	0.3138	0.3142	0.3146
8.00	0.3150	0.3154	0.3157	0.3161	0.3165	0.3169	0.3173	0.3177	0.3181	0.3185
8.10	0.3189	0.3193	0.3197	0.3201	0.3205	0.3209	0.3213	0.3217	0.3220	0.3224
8.20	0.3228	0.3232	0.3236	0.3240	0.3244	0.3248	0.3252	0.3256	0.3260	0.3264
8.30	0.3268	0.3272	0.3276	0.3280	0.3283	0.3287	0.3291	0.3295	0.3299	0.3303
8.40	0.3307	0.3311	0.3315	0.3319	0.3323	0.3327	0.3331	0.3335	0.3339	0.3343
8.50	0.3346	0.3350	0.3354	0.3358	0.3362	0.3366	0.3370	0.3374	0.3378	0.3382
8.60	0.3386	0.3390	0.3394	0.3398	0.3402	0.3406	0.3409	0.3413	0.3417	0.3421
8.70	0.3425	0.3429	0.3433	0.3437	0.3441	0.3445	0.3449	0.3453	0.3457	0.3461
8.80	0.3465	0.3469	0.3472	0.3476	0.3480	0.3484	0.3488	0.3492	0.3496	0.3500
8.90	0.3504	0.3508	0.3512	0.3516	0.3520	0.3524	0.3528	0.3531	0.3535	0.3539
9.00	0.3543	0.3547	0.3551	0.3555	0.3559	0.3563	0.3567	0.3571	0.3575	0.3579
9.10	0.3583	0.3587	0.3591	0.3594	0.3598	0.3602	0.3606	0.3610	0.3614	0.3618
9.20	0.3622	0.3626	0.3630	0.3634	0.3638	0.3642	0.3646	0.3650	0.3654	0.3657
9.30	0.3661	0.3665	0.3669	0.3673	0.3677	0.3681	0.3685	0.3689	0.3693	0.3697
9.40	0.3701	0.3705	0.3709	0.3713	0.3717	0.3720	0.3724	0.3728	0.3732	0.3736
9.50	0.3740	0.3744	0.3748	0.3752	0.3756	0.3760	0.3764	0.3768	0.3772	0.3776
9.60	0.3780	0.3783	0.3787	0.3791	0.3795	0.3799	0.3803	0.3807	0.3811	0.3815
9.70	0.3819	0.3823	0.3827	0.3831	0.3835	0.3839	0.3843	0.3846	0.3850	0.3854
9.80	0.3858	0.3862	0.3866	0.3870	0.3874	0.3878	0.3882	0.3886	0.3890	0.3894
9.90	0.3898	0.3902	0.3906	0.3909	0.3913	0.3917	0.3921	0.3925	0.3929	0.3933
10.00	0.3937	0.3941	0.3945	0.3949	0.3953	0.3957	0.3961	0.3965	0.3969	0.3972
10.10	0.3976	0.3980	0.3984	0.3988	0.3992	0.3996	0.4000	0.4004	0.4008	0.4012

mm	0.00	0.01	0.02	0.03	0.04	0.05	0.06	0.07	0.08	0.09
	Inches									
10.20	0.4016	0.4020	0.4024	0.4028	0.4031	0.4035	0.4039	0.4043	0.4047	0.4051
10.30	0.4055	0.4059	0.4063	0.4067	0.4071	0.4075	0.4079	0.4083	0.4087	0.4091
10.40	0.4094	0.4098	0.4102	0.4106	0.4110	0.4114	0.4118	0.4122	0.4126	0.4130
10.50	0.4134	0.4138	0.4142	0.4146	0.4150	0.4154	0.4157	0.4161	0.4165	0.4169
10.60	0.4173	0.4177	0.4181	0.4185	0.4189	0.4193	0.4197	0.4201	0.4205	0.4209
10.70	0.4213	0.4217	0.4220	0.4224	0.4228	0.4232	0.4236	0.4240	0.4244	0.4248
10.80	0.4252	0.4256	0.4260	0.4264	0.4268	0.4272	0.4276	0.4280	0.4283	0.4287
10.90	0.4291	0.4295	0.4299	0.4303	0.4307	0.4311	0.4315	0.4319	0.4323	0.4327
11.00	0.4331	0.4335	0.4339	0.4343	0.4346	0.4350	0.4354	0.4358	0.4362	0.4366
11.10	0.4370	0.4374	0.4378	0.4382	0.4386	0.4390	0.4394	0.4398	0.4402	0.4406
11.20	0.4409	0.4413	0.4417	0.4421	0.4425	0.4429	0.4433	0.4437	0.4441	0.4445
11.30	0.4449	0.4453	0.4457	0.4461	0.4465	0.4469	0.4472	0.4476	0.4480	0.4484
11.40	0.4488	0.4492	0.4496	0.4500	0.4504	0.4508	0.4512	0.4516	0.4520	0.4524
11.50	0.4528	0.4531	0.4535	0.4539	0.4543	0.4547	0.4551	0.4555	0.4559	0.4563
11.60	0.4567	0.4571	0.4575	0.4579	0.4583	0.4587	0.4591	0.4594	0.4598	0.4602
11.70	0.4606	0.4610	0.4614	0.4618	0.4622	0.4626	0.4630	0.4634	0.4638	0.4642
11.80	0.4646	0.4650	0.4654	0.4657	0.4661	0.4665	0.4669	0.4673	0.4677	0.4681
11.90	0.4685	0.4689	0.4693	0.4697	0.4701	0.4705	0.4709	0.4713	0.4717	0.4720
12.00	0.4724	0.4728	0.4732	0.4736	0.4740	0.4744	0.4748	0.4752	0.4756	0.4760
12.10	0.4764	0.4768	0.4772	0.4776	0.4780	0.4783	0.4787	0.4791	0.4795	0.4799
12.20	0.4803	0.4807	0.4811	0.4815	0.4819	0.4823	0.4827	0.4831	0.4835	0.4839
12.30	0.4843	0.4846	0.4850	0.4854	0.4858	0.4862	0.4866	0.4870	0.4874	0.4878
12.40	0.4882	0.4886	0.4890	0.4894	0.4898	0.4902	0.4906	0.4909	0.4913	0.4917
12.50	0.4921	0.4925	0.4929	0.4933	0.4937	0.4941	0.4945	0.4949	0.4953	0.4957
12.60	0.4961	0.4965	0.4969	0.4972	0.4976	0.4980	0.4984	0.4988	0.4992	0.4996
12.70	0.5000	0.5004	0.5008	0.5012	0.5016	0.5020	0.5024	0.5028	0.5031	0.5035
12.80	0.5039	0.5043	0.5047	0.5051	0.5055	0.5059	0.5063	0.5067	0.5071	0.5075
12.90	0.5079	0.5083	0.5087	0.5091	0.5094	0.5098	0.5102	0.5106	0.5110	0.5114
13.00	0.5118	0.5122	0.5126	0.5130	0.5134	0.5138	0.5142	0.5146	0.5150	0.5154
13.10	0.5157	0.5161	0.5165	0.5169	0.5173	0.5177	0.5181	0.5185	0.5189	0.5193
13.20	0.5197	0.5201	0.5205	0.5209	0.5213	0.5217	0.5220	0.5224	0.5228	0.5232
13.30	0.5236	0.5240	0.5244	0.5248	0.5252	0.5256	0.5260	0.5264	0.5268	0.5272
13.40	0.5276	0.5280	0.5283	0.5287	0.5291	0.5295	0.5299	0.5303	0.5307	0.5311
13.50	0.5315	0.5319	0.5323	0.5327	0.5331	0.5335	0.5339	0.5343	0.5346	0.5350
13.60	0.5354	0.5358	0.5362	0.5366	0.5370	0.5374	0.5378	0.5382	0.5386	0.5390
13.70	0.5394	0.5398	0.5402	0.5406	0.5409	0.5413	0.5417	0.5421	0.5425	0.5429
13.80	0.5433	0.5437	0.5441	0.5445	0.5449	0.5453	0.5457	0.5461	0.5465	0.5469
13.90	0.5472	0.5476	0.5480	0.5484	0.5488	0.5492	0.5496	0.5500	0.5504	0.5508

mm	0.00	0.01	0.02	0.03	0.04	0.05	0.06	0.07	0.08	0.09
	Inches									
14.00	0.5512	0.5516	0.5520	0.5524	0.5528	0.5531	0.5535	0.5539	0.5543	0.5547
14.10	0.5551	0.5555	0.5559	0.5563	0.5567	0.5571	0.5575	0.5579	0.5583	0.5587
14.20	0.5591	0.5594	0.5598	0.5602	0.5606	0.5610	0.5614	0.5618	0.5622	0.5626
14.30	0.5630	0.5634	0.5638	0.5642	0.5646	0.5650	0.5654	0.5657	0.5661	0.5665
14.40	0.5669	0.5673	0.5677	0.5681	0.5685	0.5689	0.5693	0.5697	0.5701	0.5705
14.50	0.5709	0.5713	0.5717	0.5720	0.5724	0.5728	0.5732	0.5736	0.5740	0.5744
14.60	0.5748	0.5752	0.5756	0.5760	0.5764	0.5768	0.5772	0.5776	0.5780	0.5783
14.70	0.5787	0.5791	0.5795	0.5799	0.5803	0.5807	0.5811	0.5815	0.5819	0.5823
14.80	0.5827	0.5831	0.5835	0.5839	0.5843	0.5846	0.5850	0.5854	0.5858	0.5862
14.90	0.5866	0.5870	0.5874	0.5878	0.5882	0.5886	0.5890	0.5894	0.5898	0.5902
15.00	0.5906	0.5909	0.5913	0.5917	0.5921	0.5925	0.5929	0.5933	0.5937	0.5941
15.10	0.5945	0.5949	0.5953	0.5957	0.5961	0.5965	0.5969	0.5972	0.5976	0.5980
15.20	0.5984	0.5988	0.5992	0.5996	0.6000	0.6004	0.6008	0.6012	0.6016	0.6020
15.30	0.6024	0.6028	0.6031	0.6035	0.6039	0.6043	0.6047	0.6051	0.6055	0.6059
15.40	0.6063	0.6067	0.6071	0.6075	0.6079	0.6083	0.6087	0.6091	0.6094	0.6098
15.50	0.6102	0.6106	0.6110	0.6114	0.6118	0.6122	0.6126	0.6130	0.6134	0.6138
15.60	0.6142	0.6146	0.6150	0.6154	0.6157	0.6161	0.6165	0.6169	0.6173	0.6177
15.70	0.6181	0.6185	0.6189	0.6193	0.6197	0.6201	0.6205	0.6209	0.6213	0.6217
15.80	0.6220	0.6224	0.6228	0.6232	0.6236	0.6240	0.6244	0.6248	0.6252	0.6256
15.90	0.6260	0.6264	0.6268	0.6272	0.6276	0.6280	0.6283	0.6287	0.6291	0.6295
16.00	0.6299	0.6303	0.6307	0.6311	0.6315	0.6319	0.6323	0.6327	0.6331	0.6335
16.10	0.6339	0.6343	0.6346	0.6350	0.6354	0.6358	0.6362	0.6366	0.6370	0.6374
16.20	0.6378	0.6382	0.6386	0.6390	0.6394	0.6398	0.6402	0.6406	0.6409	0.6413
16.30	0.6417	0.6421	0.6425	0.6429	0.6433	0.6437	0.6441	0.6445	0.6449	0.6453
16.40	0.6457	0.6461	0.6465	0.6469	0.6472	0.6476	0.6480	0.6484	0.6488	0.6492
16.50	0.6496	0.6500	0.6504	0.6508	0.6512	0.6516	0.6520	0.6524	0.6528	0.6531
16.60	0.6535	0.6539	0.6543	0.6547	0.6551	0.6555	0.6559	0.6563	0.6567	0.6571
16.70	0.6575	0.6579	0.6583	0.6587	0.6591	0.6594	0.6598	0.6602	0.6606	0.6610
16.80	0.6614	0.6618	0.6622	0.6626	0.6630	0.6634	0.6638	0.6642	0.6646	0.6650
16.90	0.6654	0.6657	0.6661	0.6665	0.6669	0.6673	0.6677	0.6681	0.6685	0.6689
17.00	0.6693	0.6697	0.6701	0.6705	0.6709	0.6713	0.6717	0.6720	0.6724	0.6728
17.10	0.6732	0.6736	0.6740	0.6744	0.6748	0.6752	0.6756	0.6760	0.6764	0.6768
17.20	0.6772	0.6776	0.6780	0.6783	0.6787	0.6791	0.6795	0.6799	0.6803	0.6807
17.30	0.6811	0.6815	0.6819	0.6823	0.6827	0.6831	0.6835	0.6839	0.6843	0.6846
17.40	0.6850	0.6854	0.6858	0.6862	0.6866	0.6870	0.6874	0.6878	0.6882	0.6886
17.50	0.6890	0.6894	0.6898	0.6902	0.6906	0.6909	0.6913	0.6917	0.6921	0.6925
17.60	0.6929	0.6933	0.6937	0.6941	0.6945	0.6949	0.6953	0.6957	0.6961	0.6965

mm	0.00	0.01	0.02	0.03	0.04	0.05	0.06	0.07	0.08	0.09
	Inches									
17.70	0.6969	0.6972	0.6976	0.6980	0.6984	0.6988	0.6992	0.6996	0.7000	0.7004
17.80	0.7008	0.7012	0.7016	0.7020	0.7024	0.7028	0.7031	0.7035	0.7039	0.7043
17.90	0.7047	0.7051	0.7055	0.7059	0.7063	0.7067	0.7071	0.7075	0.7079	0.7083
18.00	0.7087	0.7091	0.7094	0.7098	0.7102	0.7106	0.7110	0.7114	0.7118	0.7122
18.10	0.7126	0.7130	0.7134	0.7138	0.7142	0.7146	0.7150	0.7154	0.7157	0.7161
18.20	0.7165	0.7169	0.7173	0.7177	0.7181	0.7185	0.7189	0.7193	0.7197	0.7201
18.30	0.7205	0.7209	0.7213	0.7217	0.7220	0.7224	0.7228	0.7232	0.7236	0.7240
18.40	0.7244	0.7248	0.7252	0.7256	0.7260	0.7264	0.7268	0.7272	0.7276	0.7280
18.50	0.7283	0.7287	0.7291	0.7295	0.7299	0.7303	0.7307	0.7311	0.7315	0.7319
18.60	0.7323	0.7327	0.7331	0.7335	0.7339	0.7343	0.7346	0.7350	0.7354	0.7358
18.70	0.7362	0.7366	0.7370	0.7374	0.7378	0.7382	0.7386	0.7390	0.7394	0.7398
18.80	0.7402	0.7406	0.7409	0.7413	0.7417	0.7421	0.7425	0.7429	0.7433	0.7437
18.90	0.7441	0.7445	0.7449	0.7453	0.7457	0.7461	0.7465	0.7469	0.7472	0.7476
19.00	0.7480	0.7484	0.7488	0.7492	0.7496	0.7500	0.7504	0.7508	0.7512	0.7516
19.10	0.7520	0.7524	0.7528	0.7531	0.7535	0.7539	0.7543	0.7547	0.7551	0.7555
19.20	0.7559	0.7563	0.7567	0.7571	0.7575	0.7579	0.7583	0.7587	0.7591	0.7594
19.30	0.7598	0.7602	0.7606	0.7610	0.7614	0.7618	0.7622	0.7626	0.7630	0.7634
19.40	0.7638	0.7642	0.7646	0.7650	0.7654	0.7657	0.7661	0.7665	0.7669	0.7673
19.50	0.7677	0.7681	0.7685	0.7689	0.7693	0.7697	0.7701	0.7705	0.7709	0.7713
19.60	0.7717	0.7720	0.7724	0.7728	0.7732	0.7736	0.7740	0.7744	0.7748	0.7752
19.70	0.7756	0.7760	0.7764	0.7768	0.7772	0.7776	0.7780	0.7783	0.7787	0.7791
19.80	0.7795	0.7799	0.7803	0.7807	0.7811	0.7815	0.7819	0.7823	0.7827	0.7831
19.90	0.7835	0.7839	0.7843	0.7846	0.7850	0.7854	0.7858	0.7862	0.7866	0.7870
20.00	0.7874	0.7878	0.7882	0.7886	0.7890	0.7894	0.7898	0.7902	0.7906	0.7909
20.10	0.7913	0.7917	0.7921	0.7925	0.7929	0.7933	0.7937	0.7941	0.7945	0.7949
20.20	0.7953	0.7957	0.7961	0.7965	0.7969	0.7972	0.7976	0.7980	0.7984	0.7988
20.30	0.7992	0.7996	0.8000	0.8004	0.8008	0.8012	0.8016	0.8020	0.8024	0.8028
20.40	0.8031	0.8035	0.8039	0.8043	0.8047	0.8051	0.8055	0.8059	0.8063	0.8067
20.50	0.8071	0.8075	0.8079	0.8083	0.8087	0.8091	0.8094	0.8098	0.8102	0.8106
20.60	0.8110	0.8114	0.8118	0.8122	0.8126	0.8130	0.8134	0.8138	0.8142	0.8146
20.70	0.8150	0.8154	0.8157	0.8161	0.8165	0.8169	0.8173	0.8177	0.8181	0.8185
20.80	0.8189	0.8193	0.8197	0.8201	0.8205	0.8209	0.8213	0.8217	0.8220	0.8224
20.90	0.8228	0.8232	0.8236	0.8240	0.8244	0.8248	0.8252	0.8256	0.8260	0.8264
21.00	0.8268	0.8272	0.8276	0.8280	0.8283	0.8287	0.8291	0.8295	0.8299	0.8303
21.10	0.8307	0.8311	0.8315	0.8319	0.8323	0.8327	0.8331	0.8335	0.8339	0.8343
21.20	0.8346	0.8350	0.8354	0.8358	0.8362	0.8366	0.8370	0.8374	0.8378	0.8382
21.30	0.8386	0.8390	0.8394	0.8398	0.8402	0.8406	0.8409	0.8413	0.8417	0.8421
21.40	0.8425	0.8429	0.8433	0.8437	0.8441	0.8445	0.8449	0.8453	0.8457	0.8461
21.50	0.8465	0.8469	0.8472	0.8476	0.8480	0.8484	0.8488	0.8492	0.8496	0.8500

mm	0.00	0.01	0.02	0.03	0.04	0.05	0.06	0.07	0.08	0.09
	Inches									
21.60	0.8504	0.8508	0.8512	0.8516	0.8520	0.8524	0.8528	0.8531	0.8535	0.8539
21.70	0.8543	0.8547	0.8551	0.8555	0.8559	0.8563	0.8567	0.8571	0.8575	0.8579
21.80	0.8583	0.8587	0.8591	0.8594	0.8598	0.8602	0.8606	0.8610	0.8614	0.8618
21.90	0.8622	0.8626	0.8630	0.8634	0.8638	0.8642	0.8646	0.8650	0.8654	0.8657
22.00	0.8661	0.8665	0.8669	0.8673	0.8677	0.8681	0.8685	0.8689	0.8693	0.8697
22.10	0.8701	0.8705	0.8709	0.8713	0.8717	0.8720	0.8724	0.8728	0.8732	0.8736
22.20	0.8740	0.8744	0.8748	0.8752	0.8756	0.8760	0.8764	0.8768	0.8772	0.8776
22.30	0.8780	0.8783	0.8787	0.8791	0.8795	0.8799	0.8803	0.8807	0.8811	0.8815
22.40	0.8819	0.8823	0.8827	0.8831	0.8835	0.8839	0.8843	0.8846	0.8850	0.8854
22.50	0.8858	0.8862	0.8866	0.8870	0.8874	0.8878	0.8882	0.8886	0.8890	0.8894
22.60	0.8898	0.8902	0.8906	0.8909	0.8913	0.8917	0.8921	0.8925	0.8929	0.8933
22.70	0.8937	0.8941	0.8945	0.8949	0.8953	0.8957	0.8961	0.8965	0.8969	0.8972
22.80	0.8976	0.8980	0.8984	0.8988	0.8992	0.8996	0.9000	0.9004	0.9008	0.9012
22.90	0.9016	0.9020	0.9024	0.9028	0.9031	0.9035	0.9039	0.9043	0.9047	0.9051
23.00	0.9055	0.9059	0.9063	0.9067	0.9071	0.9075	0.9079	0.9083	0.9087	0.9091
23.10	0.9094	0.9098	0.9102	0.9106	0.9110	0.9114	0.9118	0.9122	0.9126	0.9130
23.20	0.9134	0.9138	0.9142	0.9146	0.9150	0.9154	0.9157	0.9161	0.9165	0.9169
23.30	0.9173	0.9177	0.9181	0.9185	0.9189	0.9193	0.9197	0.9201	0.9205	0.9209
23.40	0.9213	0.9217	0.9220	0.9224	0.9228	0.9232	0.9236	0.9240	0.9244	0.9248
23.50	0.9252	0.9256	0.9260	0.9264	0.9268	0.9272	0.9276	0.9280	0.9283	0.9287
23.60	0.9291	0.9295	0.9299	0.9303	0.9307	0.9311	0.9315	0.9319	0.9323	0.9327
23.70	0.9331	0.9335	0.9339	0.9343	0.9346	0.9350	0.9354	0.9358	0.9362	0.9366
23.80	0.9370	0.9374	0.9378	0.9382	0.9386	0.9390	0.9394	0.9398	0.9402	0.9406
23.90	0.9409	0.9413	0.9417	0.9421	0.9425	0.9429	0.9433	0.9437	0.9441	0.9445
24.00	0.9449	0.9453	0.9457	0.9461	0.9465	0.9469	0.9472	0.9476	0.9480	0.9484
24.10	0.9488	0.9492	0.9496	0.9500	0.9504	0.9508	0.9512	0.9516	0.9520	0.9524
24.20	0.9528	0.9531	0.9535	0.9539	0.9543	0.9547	0.9551	0.9555	0.9559	0.9563
24.30	0.9567	0.9571	0.9575	0.9579	0.9583	0.9587	0.9591	0.9594	0.9598	0.9602
24.40	0.9606	0.9610	0.9614	0.9618	0.9622	0.9626	0.9630	0.9634	0.9638	0.9642
24.50	0.9646	0.9650	0.9654	0.9657	0.9661	0.9665	0.9669	0.9673	0.9677	0.9681
24.60	0.9685	0.9689	0.9693	0.9697	0.9701	0.9705	0.9709	0.9713	0.9717	0.9720
24.70	0.9724	0.9728	0.9732	0.9736	0.9740	0.9744	0.9748	0.9752	0.9756	0.9760
24.80	0.9764	0.9768	0.9772	0.9776	0.9780	0.9783	0.9787	0.9791	0.9795	0.9799
24.90	0.9803	0.9807	0.9811	0.9815	0.9819	0.9823	0.9827	0.9831	0.9835	0.9839
25.00	0.9843	0.9846	0.9850	0.9854	0.9858	0.9862	0.9866	0.9870	0.9874	0.9878
25.10	0.9882	0.9886	0.9890	0.9894	0.9898	0.9902	0.9906	0.9909	0.9913	0.9917
25.20	0.9921	0.9925	0.9929	0.9933	0.9937	0.9941	0.9945	0.9949	0.9953	0.9957
25.30	0.9961	0.9965	0.9969	0.9972	0.9976	0.9980	0.9984	0.9988	0.9992	0.9996
25.40	1.0000	1.0004	1.0008	1.0012	1.0016	1.0020	1.0024	1.0028	1.0031	1.0035

Table 27.2. Millimeters to inches: 0 to 300mm in 1mm stages

mm	0	1	2	3	4	5	6	7	8	9
	Inches									
0	0.000	0.039	0.079	0.118	0.157	0.197	0.236	0.276	0.315	0.354
10	0.394	0.433	0.472	0.512	0.551	0.591	0.630	0.669	0.709	0.748
20	0.787	0.827	0.866	0.906	0.945	0.984	1.024	1.063	1.102	1.142
30	1.181	1.220	1.260	1.299	1.339	1.378	1.417	1.457	1.496	1.535
40	1.575	1.614	1.654	1.693	1.732	1.772	1.811	1.850	1.890	1.929
50	1.969	2.008	2.047	2.087	2.126	2.165	2.205	2.244	2.283	2.323
60	2.362	2.402	2.441	2.480	2.520	2.559	2.598	2.638	2.677	2.717
70	2.756	2.795	2.835	2.874	2.913	2.953	2.992	3.031	3.071	3.110
80	3.150	3.189	3.228	3.268	3.307	3.346	3.386	3.425	3.465	3.504
90	3.543	3.583	3.622	3.661	3.701	3.740	3.780	3.819	3.858	3.898
100	3.937	3.976	4.016	4.055	4.094	4.134	4.173	4.213	4.252	4.291
110	4.331	4.370	4.409	4.449	4.488	4.528	4.567	4.606	4.646	4.685
120	4.724	4.764	4.803	4.843	4.882	4.921	4.961	5.000	5.039	5.079
130	5.118	5.157	5.197	5.236	5.276	5.315	5.354	5.394	5.433	5.472
140	5.512	5.551	5.591	5.630	5.669	5.709	5.748	5.787	5.827	5.866
150	5.906	5.945	5.984	6.024	6.063	6.102	6.142	6.181	6.220	6.260
160	6.299	6.339	6.378	6.417	6.457	6.496	6.535	6.575	6.614	6.654
170	6.693	6.732	6.772	6.811	6.850	6.890	6.929	6.969	7.008	7.047
180	7.087	7.126	7.165	7.205	7.244	7.283	7.323	7.362	7.402	7.441
190	7.480	7.520	7.559	7.598	7.638	7.677	7.717	7.756	7.795	7.835
200	7.874	7.913	7.953	7.992	8.031	8.071	8.110	8.150	8.189	8.228
210	8.268	8.307	8.346	8.386	8.425	8.465	8.504	8.543	8.583	8.622
220	8.661	8.701	8.740	8.780	8.819	8.858	8.898	8.937	8.976	9.016
230	9.055	9.094	9.134	9.173	9.213	9.252	9.291	9.331	9.370	9.409
240	9.449	9.488	9.528	9.567	9.606	9.646	9.685	9.724	9.764	9.803
250	9.843	9.882	9.921	9.961	10.000	10.039	10.079	10.118	10.157	10.197
260	10.236	10.276	10.315	10.354	10.394	10.433	10.472	10.512	10.551	10.591
270	10.630	10.669	10.709	10.748	10.787	10.827	10.866	10.906	10.945	10.984
280	11.024	11.063	11.102	11.142	11.181	11.220	11.260	11.299	11.339	11.378
290	11.417	11.457	11.496	11.535	11.575	11.614	11.654	11.693	11.732	11.772
300	11.811	11.850	11.890	11.929	11.969	12.008	12.047	12.087	12.126	12.165

Table 27.3. Inches to millimeters: 0 to 1in in 1 thou. stages

Inches	0.000	0.001	0.002	0.003	0.004	0.005	0.006	0.007	0.008	0.009
	Millimeters									
0	0.00	0.03	0.05	0.08	0.10	0.13	0.15	0.18	0.20	0.23
0.010	0.25	0.28	0.30	0.33	0.36	0.38	0.41	0.43	0.46	0.48
0.020	0.51	0.53	0.56	0.58	0.61	0.63	0.66	0.69	0.71	0.74
0.030	0.76	0.79	0.81	0.84	0.86	0.89	0.91	0.94	0.97	0.99
0.040	1.02	1.04	1.07	1.09	1.12	1.14	1.17	1.19	1.22	1.24
0.050	1.27	1.30	1.32	1.35	1.37	1.40	1.42	1.45	1.47	1.50
0.060	1.52	1.55	1.57	1.60	1.63	1.65	1.68	1.70	1.73	1.75
0.070	1.78	1.80	1.83	1.85	1.88	1.90	1.93	1.96	1.98	2.01
0.080	2.03	2.06	2.08	2.11	2.13	2.16	2.18	2.21	2.24	2.26
0.090	2.29	2.31	2.34	2.36	2.39	2.41	2.44	2.46	2.49	2.51
0.100	2.54	2.57	2.59	2.62	2.64	2.67	2.69	2.72	2.74	2.77
0.110	2.79	2.82	2.84	2.87	2.90	2.92	2.95	2.97	3.00	3.02
0.120	3.05	3.07	3.10	3.12	3.15	3.17	3.20	3.23	3.25	3.28
0.130	3.30	3.33	3.35	3.38	3.40	3.43	3.45	3.48	3.51	3.53
0.140	3.56	3.58	3.61	3.63	3.66	3.68	3.71	3.73	3.76	3.78
0.150	3.81	3.84	3.86	3.89	3.91	3.94	3.96	3.99	4.01	4.04
0.160	4.06	4.09	4.11	4.14	4.17	4.19	4.22	4.24	4.27	4.29
0.170	4.32	4.34	4.37	4.39	4.42	4.44	4.47	4.50	4.52	4.55
0.180	4.57	4.60	4.62	4.65	4.67	4.70	4.72	4.75	4.78	4.80
0.190	4.83	4.85	4.88	4.90	4.93	4.95	4.98	5.00	5.03	5.05
0.200	5.08	5.11	5.13	5.16	5.18	5.21	5.23	5.26	5.28	5.31
0.210	5.33	5.36	5.38	5.41	5.44	5.46	5.49	5.51	5.54	5.56
0.220	5.59	5.61	5.64	5.66	5.69	5.72	5.74	5.77	5.79	5.82
0.230	5.84	5.87	5.89	5.92	5.94	5.97	5.99	6.02	6.05	6.07
0.240	6.10	6.12	6.15	6.17	6.20	6.22	6.25	6.27	6.30	6.32
0.250	6.35	6.38	6.40	6.43	6.45	6.48	6.50	6.53	6.55	6.58
0.260	6.60	6.63	6.65	6.68	6.71	6.73	6.76	6.78	6.81	6.83
0.270	6.86	6.88	6.91	6.93	6.96	6.99	7.01	7.04	7.06	7.09
0.280	7.11	7.14	7.16	7.19	7.21	7.24	7.26	7.29	7.32	7.34
0.290	7.37	7.39	7.42	7.44	7.47	7.49	7.52	7.54	7.57	7.59
0.300	7.62	7.65	7.67	7.70	7.72	7.75	7.77	7.80	7.82	7.85
0.310	7.87	7.90	7.92	7.95	7.98	8.00	8.03	8.05	8.08	8.10
0.320	8.13	8.15	8.18	8.20	8.23	8.26	8.28	8.31	8.33	8.36
0.330	8.38	8.41	8.43	8.46	8.48	8.51	8.53	8.56	8.59	8.61
0.340	8.64	8.66	8.69	8.71	8.74	8.76	8.79	8.81	8.84	8.86
0.350	8.89	8.92	8.94	8.97	8.99	9.02	9.04	9.07	9.09	9.12

Inches	0.000	0.001	0.002	0.003	0.004	0.005	0.006	0.007	0.008	0.009
	Millimeters									
0.360	9.14	9.17	9.19	9.22	9.25	9.27	9.30	9.32	9.35	9.37
0.370	9.40	9.42	9.45	9.47	9.50	9.53	9.55	9.58	9.60	9.63
0.380	9.65	9.68	9.70	9.73	9.75	9.78	9.80	9.83	9.86	9.88
0.390	9.91	9.93	9.96	9.98	10.01	10.03	10.06	10.08	10.11	10.13
0.400	10.16	10.19	10.21	10.24	10.26	10.29	10.31	10.34	10.36	10.39
0.410	10.41	10.44	10.46	10.49	10.52	10.54	10.57	10.59	10.62	10.64
0.420	10.67	10.69	10.72	10.74	10.77	10.80	10.82	10.85	10.87	10.90
0.430	10.92	10.95	10.97	11.00	11.02	11.05	11.07	11.10	11.13	11.15
0.440	11.18	11.20	11.23	11.25	11.28	11.30	11.33	11.35	11.38	11.40
0.450	11.43	11.46	11.48	11.51	11.53	11.56	11.58	11.61	11.63	11.66
0.460	11.68	11.71	11.73	11.76	11.79	11.81	11.84	11.86	11.89	11.91
0.470	11.94	11.96	11.99	12.01	12.04	12.07	12.09	12.12	12.14	12.17
0.480	12.19	12.22	12.24	12.27	12.29	12.32	12.34	12.37	12.40	12.42
0.490	12.45	12.47	12.50	12.52	12.55	12.57	12.60	12.62	12.65	12.67
0.500	12.70	12.73	12.75	12.78	12.80	12.83	12.85	12.88	12.90	12.93
0.510	12.95	12.98	13.00	13.03	13.06	13.08	13.11	13.13	13.16	13.18
0.520	13.21	13.23	13.26	13.28	13.31	13.34	13.36	13.39	13.41	13.44
0.530	13.46	13.49	13.51	13.54	13.56	13.59	13.61	13.64	13.67	13.69
0.540	13.72	13.74	13.77	13.79	13.82	13.84	13.87	13.89	13.92	13.94
0.550	13.97	14.00	14.02	14.05	14.07	14.10	14.12	14.15	14.17	14.20
0.560	14.22	14.25	14.27	14.30	14.33	14.35	14.38	14.40	14.43	14.45
0.570	14.48	14.50	14.53	14.55	14.58	14.61	14.63	14.66	14.68	14.71
0.580	14.73	14.76	14.78	14.81	14.83	14.86	14.88	14.91	14.94	14.96
0.590	14.99	15.01	15.04	15.06	15.09	15.11	15.14	15.16	15.19	15.21
0.600	15.24	15.27	15.29	15.32	15.34	15.37	15.39	15.42	15.44	15.47
0.610	15.49	15.52	15.54	15.57	15.60	15.62	15.65	15.67	15.70	15.72
0.620	15.75	15.77	15.80	15.82	15.85	15.88	15.90	15.93	15.95	15.98
0.630	16.00	16.03	16.05	16.08	16.10	16.13	16.15	16.18	16.21	16.23
0.640	16.26	16.28	16.31	16.33	16.36	16.38	16.41	16.43	16.46	16.48
0.650	16.51	16.54	16.56	16.59	16.61	16.64	16.66	16.69	16.71	16.74
0.660	16.76	16.79	16.81	16.84	16.87	16.89	16.92	16.94	16.97	16.99
0.670	17.02	17.04	17.07	17.09	17.12	17.15	17.17	17.20	17.22	17.25
0.680	17.27	17.30	17.32	17.35	17.37	17.40	17.42	17.45	17.48	17.50
0.690	17.53	17.55	17.58	17.60	17.63	17.65	17.68	17.70	17.73	17.75
0.700	17.78	17.81	17.83	17.86	17.88	17.91	17.93	17.96	17.98	18.01
0.710	18.03	18.06	18.08	18.11	18.14	18.16	18.19	18.21	18.24	18.26
0.720	18.29	18.31	18.34	18.36	18.39	18.42	18.44	18.47	18.49	18.52
0.730	18.54	18.57	18.59	18.62	18.64	18.67	18.69	18.72	18.75	18.77

Inches	0.000	0.001	0.002	0.003	0.004	0.005	0.006	0.007	0.008	0.009
	Millimeters									
0.740	18.80	18.82	18.85	18.87	18.90	18.92	18.95	18.97	19.00	19.02
0.750	19.05	19.08	19.10	19.13	19.15	19.18	19.20	19.23	19.25	19.28
0.760	19.30	19.33	19.35	19.38	19.41	19.43	19.46	19.48	19.51	19.53
0.770	19.56	19.58	19.61	19.63	19.66	19.69	19.71	19.74	19.76	19.79
0.780	19.81	19.84	19.86	19.89	19.91	19.94	19.96	19.99	20.02	20.04
0.790	20.07	20.09	20.12	20.14	20.17	20.19	20.22	20.24	20.27	20.29
0.800	20.32	20.35	20.37	20.40	20.42	20.45	20.47	20.50	20.52	20.55
0.810	20.57	20.60	20.62	20.65	20.68	20.70	20.73	20.75	20.78	20.80
0.820	20.83	20.85	20.88	20.90	20.93	20.96	20.98	21.01	21.03	21.06
0.830	21.08	21.11	21.13	21.16	21.18	21.21	21.23	21.26	21.29	21.31
0.840	21.34	21.36	21.39	21.41	21.44	21.46	21.49	21.51	21.54	21.56
0.850	21.59	21.62	21.64	21.67	21.69	21.72	21.74	21.77	21.79	21.82
0.860	21.84	21.87	21.89	21.92	21.95	21.97	22.00	22.02	22.05	22.07
0.870	22.10	22.12	22.15	22.17	22.20	22.23	22.25	22.28	22.30	22.33
0.880	22.35	22.38	22.40	22.43	22.45	22.48	22.50	22.53	22.56	22.58
0.890	22.61	22.63	22.66	22.68	22.71	22.73	22.76	22.78	22.81	22.83
0.900	22.86	22.89	22.91	22.94	22.96	22.99	23.01	23.04	23.06	23.09
0.910	23.11	23.14	23.16	23.19	23.22	23.24	23.27	23.29	23.32	23.34
0.920	23.37	23.39	23.42	23.44	23.47	23.50	23.52	23.55	23.57	23.60
0.930	23.62	23.65	23.67	23.70	23.72	23.75	23.77	23.80	23.83	23.85
0.940	23.88	23.90	23.93	23.95	23.98	24.00	24.03	24.05	24.08	24.10
0.950	24.13	24.16	24.18	24.21	24.23	24.26	24.28	24.31	24.33	24.36
0.960	24.38	24.41	24.43	24.46	24.49	24.51	24.54	24.56	24.59	24.61
0.970	24.64	24.66	24.69	24.71	24.74	24.77	24.79	24.82	24.84	24.87
0.980	24.89	24.92	24.94	24.97	24.99	25.02	25.04	25.07	25.10	25.12
0.990	25.15	25.17	25.20	25.22	25.25	25.27	25.30	25.32	25.35	25.37

Table 27.4. Inches to millimeters: 0 to 190in in 1in intervals

Inches	0	1	2	3	4	5	6	7	8	9
	Millimeters									
0	0	25.4	50.8	76.2	101.6	127.0	152.4	177.8	203.2	228.6
10	254.0	279.4	304.8	330.2	355.6	381.0	406.4	431.8	457.2	482.6
20	508.0	533.4	558.8	584.2	609.6	635.0	660.4	685.8	711.2	736.6
30	762.0	787.4	812.8	838.2	863.6	889.0	914.4	939.8	965.2	990.6
40	1016.0	1041.4	1066.8	1092.2	1117.6	1143.0	1168.4	1193.8	1219.2	1244.6
50	1270.0	1295.4	1320.8	1346.2	1371.6	1397.0	1422.4	1447.8	1473.2	1498.6
60	1524.0	1549.4	1574.8	1600.2	1625.6	1651.0	1676.4	1701.8	1727.2	1752.6
70	1778.0	1803.4	1828.8	1854.2	1879.6	1905.0	1930.4	1955.8	1981.2	2006.6
80	2032.0	2057.4	2082.8	2108.2	2133.6	2159.0	2184.4	2209.8	2235.2	2260.6
90	2286.0	2311.4	2336.8	2362.2	2387.6	2413.0	2438.4	2463.8	2489.2	2514.6
100	2540.0	2565.4	2590.8	2616.2	2641.6	2667.0	2692.4	2717.8	2743.2	2768.6
110	2794.0	2819.4	2844.8	2870.2	2895.6	2921.0	2946.4	2971.8	2997.2	3022.6
120	3048.0	3073.4	3098.8	3124.2	3149.6	3175.0	3200.4	3225.8	3251.2	3276.6
130	3302.0	3327.4	3352.8	3378.2	3403.6	3429.0	3454.4	3479.8	3505.2	3530.6
140	3556.0	3581.4	3606.8	3632.2	3657.6	3683.0	3708.4	3733.8	3759.2	3784.6
150	3810.0	3835.4	3860.8	3886.2	3911.6	3937.0	3962.4	3987.8	4013.2	4038.6
160	4064.0	4089.4	4114.8	4140.2	4165.6	4191.0	4216.4	4241.8	4267.2	4292.6
170	4318.0	4343.4	4368.8	4394.2	4419.6	4445.0	4470.4	4495.8	4521.2	4546.6
180	4572.0	4597.4	4622.8	4648.2	4673.6	4699.0	4724.4	4749.8	4775.2	4800.6
190	4826.0	4851.4	4876.8	4902.2	4927.6	4953.0	4978.4	5003.8	5029.2	5054.6

**Table 27.5. Inches (fractions) to inches (decimal) and millimeters:
0 to 1in in 1/64 inch stages**

Inches		Metric
Fractions	Decimal	mm
1/64	0.01563	0.3969
1/32	0.03125	0.7937
3/64	0.04688	1.1906
1/16	0.06250	1.5875
5/64	0.07813	1.9844
3/32	0.09375	2.3812
7/64	0.10938	2.7781
1/8	0.12500	3.1750
9/64	0.14063	3.5719
5/32	0.15625	3.9688
11/64	0.17188	4.3656
3/16	0.18750	4.7625
13/64	0.20313	5.1594
7/32	0.21875	5.5562
15/64	0.23438	5.9531
1/4	0.25000	6.3500
17/64	0.26563	6.7469
9/32	0.28125	7.1437
19/64	0.29688	7.5406
5/16	0.31250	7.9375
21/64	0.32813	8.3344
11/32	0.34375	8.7312
23/64	0.35938	9.1281
3/8	0.37500	9.5250
25/64	0.39063	9.9219
13/32	0.40625	10.3187
27/64	0.42188	10.7156
7/16	0.43750	11.1125
29/64	0.45313	11.5094
15/32	0.46875	11.9063
31/64	0.48438	12.3031
1/2	0.50000	12.7000

Inches		Metric
Fractions	Decimal	mm
33/64	0.51563	13.0969
17/32	0.53125	13.4937
35/64	0.54688	13.8906
9/16	0.56250	14.2875
37/64	0.57813	14.6844
19/32	0.59375	15.0812
39/64	0.60938	15.4781
5/8	0.62500	15.8750
41/64	0.64063	16.2719
21/32	0.65625	16.6687
43/64	0.67188	17.0656
11/16	0.68750	17.4625
45/64	0.70313	17.8594
23/32	0.71875	18.2562
47/64	0.73438	18.6531
3/4	0.75000	19.0500
49/64	0.76563	19.4469
25/32	0.78125	19.8438
51/64	0.79688	20.2406
13/16	0.81250	20.6375
53/64	0.82813	21.0344
27/32	0.84375	21.4312
55/64	0.85938	21.8281
7/8	0.87500	22.2250
57/64	0.89063	22.6219
29/32	0.90625	23.0187
59/64	0.92188	23.4156
15/16	0.93750	23.8125
61/64	0.95313	24.2094
31/32	0.96875	24.6062
63/64	0.98438	25.0031
1	1.00000	25.4000

TEMPERATURE CONVERSION

Table 27.6. Fahrenheit to Centigrade

°F	°C	°F	°C	°F	°C
-10.00	-23.33	160.00	71.11	400.00	204.44
0	-17.78	170.00	76.67		
10.00	-12.22	180.00	82.22	440.00	226.67
20.00	-6.67			480.00	248.89
30.00	-1.11	190.00	87.78	520.00	271.11
		200.00	93.33	560.00	293.33
40.00	4.44	210.00	98.89	600.00	315.56
50.00	10.00	220.00	104.44		
60.00	15.56	230.00	110.00	640.00	337.78
70.00	21.11			680.00	360.00
80.00	26.67	240.00	115.56	720.00	382.22
		250.00	121.11	760.00	404.44
90.00	32.22	260.00	126.67	800.00	426.67
100.00	37.78	280.00	137.78		
110.00	43.33	300.00	148.89	840.00	448.89
120.00	48.89			880.00	471.11
130.00	54.44	320.00	160.00	920.00	493.33
		340.00	171.11	960.00	515.56
140.00	60.00	360.00	182.22	1000.00	537.78
150.00	65.56	380.00	193.33		

Table 27.7. Centigrade to Fahrenheit

°C	°F	°C	°F	°C	°F
-20.00	-4.00	70.00	158.00	220.00	428.00
-15.00	5.00	75.00	167.00	240.00	464.00
-10.00	14.00			260.00	500.00
-5.00	23.00	80.00	176.00	280.00	536.00
0	32.00	85.00	185.00	300.00	572.00
		90.00	194.00		
5.00	41.00	95.00	203.00	320.00	608.00
10.00	50.00	100.00	212.00	340.00	644.00
15.00	59.00			360.00	680.00
20.00	68.00	110.00	230.00	380.00	716.00
25.00	77.00	120.00	248.00	400.00	752.00
		130.00	266.00		
30.00	86.00	140.00	284.00	420.00	788.00
35.00	95.00	150.00	302.00	440.00	824.00
40.00	104.00			460.00	860.00
45.00	113.00	160.00	320.00	480.00	896.00
50.00	122.00	170.00	338.00	500.00	932.00
		180.00	356.00		
55.00	131.00	190.00	374.00	520.00	968.00
60.00	140.00	200.00	392.00	540.00	1004.00
65.00	149.00			560.00	1040.00
				580.00	1076.00
				600.00	1112.00

CONVERSION FACTORS

To use the tables, multiply the value of the first unit (first column) by the conversion factor to get the value of the second unit (third column). Conversely, divide the value of the second unit by the conversion factor to get the value of the first unit.

The letter 'E' in the right-hand column indicates that the conversion factor given is exact. Some factors have exact values if taken beyond four decimal places, but most do not.

Table 27.8. Angular

degrees	60.0000	minutes	E
minutes	60.0000	seconds	E

Table 27.9. Area

acres	4046.8564	square meters	-
acres	4840.0000	square yards	E
hectares	2.4711	acres	-
hectares	10000.0000	square meters	E
hectares	11959.9005	square yards	-
square feet	144.0000	square inches	E
square inches	645.1600	square millimeters	E
square meters	1.1960	square yards	-
square meters	10.7639	square feet	-
square miles	258.9988	hectares	-
square miles	640.0000	acres	E
square yards	9.0000	square feet	E

Table 27.10. Energy

British thermal units	778.1693	foot pounds	-
British thermal units	1055.0560	joules	-
foot pounds	1.3558	joules	-
joules	1.0000	newton meters	E
joules	1.0000	watt seconds	E
kilowatt hours	3.6000	megajoules	E
kilowatt hours	3412.1282	British thermal units	-

Table 27.11. Length

fathoms	1.8288	meters	E
fathoms	6.0000	feet	E
feet	12.0000	inches	E
feet	304.8000	millimeters	E
furlongs	201.1680	meters	E
furlongs	220.0000	00 yards	E
inches	25.4000	millimeters	E
kilometers	1093.6133	yards	-
meters	1.0936	yards	-
meters	3.2808	feet	-
meters	39.3701	inches	-
miles	1.6093	kilometers	-
miles	8.0000	furlongs	E
miles	1760.0000	yards	E
nautical miles	1.1515	miles	-
nautical miles	1.8532	kilometers	-
nautical miles	6080.0000	feet	E
nautical miles, international	1852.0000	meters	E
nautical miles, international	6076.1155	feet	-
yards	3.0000	feet	E
yards	36.0000	inches	E
yards	914.4000	millimeters	E

Table 27.12. Power

foot pounds per second	1.3558	watts	-
horsepower	550.0000	foot pounds per second	E
horsepower	745.7010	watts	-
horsepower	33000.0000	foot pounds per minute	E
watts	3.4121	British thermal units per hour	-

Table 27.13. Torque

kilogram meters	7.2330	pounds feet	-
kilogram meters	9.8066	newton meters	-
kilogram meters	86.7961	pounds inches	-
pounds feet	1.3558	newton meters	-

Table 27.14. Velocity

feet per second	1.0973	kilometers per hour	-
feet per second	18.2880	meters per minute	E
kilometers per hour	16.6667	meters per minute	-
kilometers per hour	54.6807	feet per minute	-
knots	1.0000	nautical miles per hour	E
knots	1.1515	miles per hour	-
knots	1.6889	feet per second	-
knots	1.8532	kilometers per hour	-
meters per minute	3.2808	feet per minute	-
miles per hour	1.6093	kilometers per hour	-
miles per hour	26.8224	meters per minute	E
miles per hour	88.0000	feet per minute	E

Table 27.15. Volume

cubic feet	28.3168	liters	-
cubic feet	1728.0000	cubic inches	E
cubic inches	16.3871	cubic centimeters	-
cubic meters	1.3080	cubic yards	-
cubic meters	35.3147	cubic feet	-
cubic meters	1000.0000	liters	E
cubic yards	27.0000	cubic feet	E
gallons (UK)	1.2009	gallons (US)	-
gallons (UK)	4.5461	liters	-
gallons (UK)	8.0000	pints	E
gallons (UK)	277.4194	cubic inches	-
gallons (US)	3.7854	liters	-
gallons (US)	8.0000	pints (US)	E
gallons (US)	231.0000	cubic inches	E
liters	61.0237	cubic inches	-
liters	1000.0000	cubic centimeters	E
pints	20.0000	fluid ounces	E
quarts	2.0000	pints	E

Table 27.16. Weight

kilograms	2.2046	pounds	-
metric tons/tons	1000.0000	kilograms	E
metric tons/tons	2204.6200	pounds	-
ounces	28.3495	grams	-
pounds	16.0000	ounces	E
pounds	453.5924	grams	-
tons	1.0160	metric tons/ton	-
tons	1016.0481	kilograms	-
tons	2240.0000	pounds	E

LESS USED MEASURES

Table 27.17. Length

chains	4.0000	rods	E
chains	66.0000	feet	E
chains	100.0000	links	E
hands	4.0000	inches	E
rods	16.5000	feet	E
spans	9.0000	inches	E

Table 27.18. Pressure

atmospheres	30.0000	inches of mercury	-
atmospheres	14.6959	pounds per square inch	-
kilograms per square centimeter	14.2233	pounds per square inch	-

Table 27.19. Quantities

dozen	12 items	E
gross	144 items	E
quire	25 sheets*	E
ream	500 sheets*	E
score	20 items	E

Earlier standard: 24 and 480 sheets.

Table 27.20. Volume

bushels	64.0000	pints	E
pecks	16.0000	pints	E
pints	4.0000	gills	E

Table 27.21. Weight

grains	1.0000	grains Troy	E
grams	15.4323	grains	-
hundredweights	4.0000	quarters	E
hundredweights	112.0000	pounds	E
ounces	437.5000	grains	E
pounds	7000.0000	grains	E
quarters	28.0000	pounds	E
stones	14.0000	pounds	E

Table 27.22. Weight: Troy (precious metals etc.)

carats (diamonds)	3.0860	grains	-
grains Troy	1.0000	grains	E
ounces Troy	31.1035	grams	-
ounces Troy	480.0000	grains	E
pennyweights	24.0000	grains	E
pounds Troy	5760.0000	grains	E
pounds Troy	12.0000	ounces Troy	E

Table 27.23. Temperature

Degrees Centigrade to Fahrenheit	$(°C \times 9/5) + 32$	E
Degrees Fahrenheit to Centigrade	$(°F - 32) \times 5/9$	E

FURTHER INFORMATION

In a few subject areas, it has been necessary to limit the listings of data in this book, so readers requiring more details may need to refer to other sources of information.

INTERNATIONAL (ISO) AND BRITISH STANDARDS

Many of the subjects are covered in much greater detail in the International and British Standards, which, for a given subject are now equal, or very nearly so, to each other. The standards of other countries may also be based on the International Standards.

In the UK, printed copies of British Standards can be viewed in some local reference libraries though most now have access to them on line. For details of the location of your nearest library having access to the standards, contact the address below.

While British Standards cannot otherwise be viewed on line without payment of a fee, much technical information is often available within manufacturers catalogs which can be downloaded, usually in pdf format.

For further information about British Standards, contact:

British Standards Institute
12950 Worldgate Drive, Suite 800
Herndon, VA 20170
Tel. 1 800 862 4977
Fax. 1 703 437 9001

PUBLICATIONS

The following books in this series contain data of interest to the small workshop owner, and some give further details about the items covered in this book.

Backyard Foundry for Home Machinists (ISBN 978-1-56523-865-7)

Basic Lathework for Home Machinists (ISBN 978-1-56523-696-7)

Metal Lathe for Home Machinists (ISBN 978-1-56523-693-6)

The Metalworker's Workshop for Home Machinists (ISBN 978-1-56523-697-4)

Milling for Home Machinists (ISBN 978-1-56523-694-3)

The Milling Machine for Home Machinists (ISBN 978-1-56523-769-8)

Mini-Lathe for Home Machinists (ISBN 978-1-56523-695-0)

Useful Machine Shop Tools to Make for Home Shop Machinists (ISBN 978-1-56523-864-0)

Two of the best-known reference books are:

Machinery's Handbook, published by Industrial Press, New York

Kempe's Engineers Year Book, published by MG Information Services Ltd.

Both books are regularly updated and cover a very much wider range of subjects, and in greater detail, although the information is mostly beyond that applicable to the small workshop.

GLOSSARY OF ENGINEERING TERMS

The definitions given are of necessity brief but, where insufficient, should point the reader in the direction for more detailed information. Almost all the terms listed have been used consistently over the years. However, a few have been rather loosely applied and the descriptions given may not always be applicable.

A

Arbor: A machine-mounted rotating spindle on to which is fitted a milling cutter, grinding wheel or other such device. Also a spindle on to which a component is fitted for the purposes of being machined. See also Mandrel.

Acme thread: A variation on a square thread where the sides of the thread slope by a few degrees, but still have a flat top and flat bottom to the thread form.

Allen screw: A name occasionally used for cap screws. See also Cap screws.

Annealing: Softening metal by heating to a high temperature and allowing it to cool slowly. This can enable the metal to be formed by bending or beating or some other purpose. See also Tempering.

Apron: The part of a lathe which hangs down from the saddle and in front of the bed, and which carries the controls for operating the saddle.

Axial: In line with the long axis of a part. In electrical components, a part with the leads projecting from each end, as in the case of the common resistor. See also Radial.

B

Back gear: The gear chain associated with a lathe mandrel that, when brought into use, reduces each of the normal speeds available, typically by a factor of 7:1. Used when screw cutting or carrying out heavy or intermittent machining.

Back plate: An adapter plate mounted on to the rear of a lathe chuck and machined to suit the mounting arrangement of the lathe on to which it is to be fitted.

Blotter: The paper disk that goes between a grinding wheel and the metal flanges used on either side for gripping the wheel. The type of paper is important; do not use ordinary paper.

Bull gear: The gear in the back gear chain that is permanently fixed to the lathe mandrel.

Buttress thread: A thread with one upright face, as in a square thread, and one sloping face, as in a normal thread, normally at 45°.

C

Cap screw: A screw or bolt with a cylindrical head and a hexagonal recess for use with a hexagonal key for tightening the screw. See also Allen screw.

Capstan lathe: A lathe, very similar to a center lathe, but in place of the tailstock a rotary turret is fitted in which a variety of tools can be fitted and used in sequence. Using adjustable stops on the turret, saddle and cross slide, the lathe can be set-up for efficient batch production. Also known as a 'Turret lathe.'

Carriage: See Saddle.

Case hardening: The action of heating a piece of mild steel to red heat while enclosed in an appropriate compound, which gives a surface hardness. The depth of the hard surface will typically be in the order of 0.1mm, although it can be much more depending on a number of factors.

Castle nut: A nut with slots across its top face that gives a castle tower-like top. One slot, together with a hole in the screw or stud, permits a split pin to be fitted through slot and hole to lock the nut in position. Also known as a 'Slotted nut.'

Catch plate: A small plate mounted on to the lathe mandrel, which, via a slot, or a post mounted on to it, drives a workpiece mounted between centers. The workpiece is fitted with a lathe carrier (sometimes called a 'dog') to take up the drive.

Center height: The height of the lathe mandrel above the lathe bed and therefore equal to half the maximum diameter that the lathe can accommodate. Used in the UK to describe a lathe's size. See also Swing.

Changewheel: The gears that are mounted between the outer end of the lathe mandrel and the leadscrew. By altering their sizes, different threads can be cut or fine feeds obtained.

Circular pitch: See Chapter 12 re gearing.

Climb milling: See Down milling.

Combination drill: Another name for a center drill.

Compound slide: The short slide mounted on top of the lathe cross slide, more usually called the top slide. See also Compound table.

Compound table: A device with two slideways mounted at a 90° angle, sometimes adjustable. Typically the working table of a mill/drill, but most often applied to a free-standing table for use in inspection or mounted on to the table of a larger machine.

Conventional milling: See Up milling.

Cotter pin: Another name for a split pin.

Cross drilling: Drilling a radial hole through the center of a round piece of material. See also Radial.

D

Detent: A locating plunger, usually sprung-loaded, typically used for locating the positions on a dividing plate.

Dead center: Originally a tailstock mounted center that does not rotate with the workpiece and, as a result, needed to be hardened. Now applied to any solid center, as opposed to one mounted in bearings and called a live center (see Live center). Dead center is also used to define the point that is exactly central.

Dial indicator: A measuring device that has a dial read-out with a wide range, intended for actual measuring, and a plunger measuring device. See also Dial test indicator.

Dial test indicator: A measuring device with a dial read-out that has a limited range, e.g. 0 to 0.5mm. It is intended for comparative testing, concentricity, etc. Almost always has a pivoted measuring arm. See also Dial indicator.

Diametral pitch: See Chapter 12 re gearing.

Direct on-line starter: A starter that switches an electric motor on to the power source in one step, i.e. it has no means of raising the voltage from off to fully on, either gradually or in steps.

Division plate: A plate with one or more series of holes on a pitch circle diameter, each having a given number of holes and used to give movement in fixed angular steps for machining gears and similar operations.

Dog: A device that transmits the drive from the catch plate to a workpiece mounted between centers. More often called the lathe carrier. Also used as a name for devices that assist the mounting of workpieces on to a faceplate.

Dog clutch: A clutch that transmits power using a peg and slot rather than friction pads, thus ensuring that input and output are always in the same angular position.

Dowel pin: A parallel pin for fitting into close-fitting holes in two or more parts. This ensures that, when dismantled, the parts can accurately be returned to the same position on re-assembly. See also Taper pin.

Down milling: Where rotation of the cutter and movement of the workpiece are in the same direction. This should be avoided except on machines with zero backlash leadscrews, where it is often the preferred method as a better finish can be achieved. Frequently known as 'Climb milling'. See also Up milling.

Draw bar: A bar threaded into the small end of the taper of a milling chuck, or similar, which passes through a milling machine spindle being held with a nut at the other end. This pulls the chuck taper into the machine spindle and prevents it from working loose while in use.

Draw filing: Using a file with a sideways motion rather than in line with its length. This is done with the file held across the body, with one hand on the handle and the other on the file's outer end and moving it towards and away from the body. This gives a good finish and is best carried out with a fine single-cut file.

Dressing: The act of removing the outer surface of a grinding wheel using a dressing device, usually a diamond mounted in a holder. This is done initially to make the wheel run true, and subsequently, after much use, to open up the surface to keep it cutting freely.

Drift: A wedge-shaped piece of steel used to eject Morse taper drills and drill chucks from the drilling machine spindle. An aperture is provided in the drilling machine quill for this purpose.

Drill rod: An American term for steel very similar to Silver steel. See also Silver steel.

Driver: The gear in a pair of gears that drives the other gear.

Driven: The gear in a pair of gears that is driven by the other gear

Double cut: A file made with two sets of cutting edges, one a mirror image of the other. This produces a file with a diamond-shaped tooth form.

F

Fabricated: Making a single part from a number of separate items, usually where a part is built up from separate items by welding, soldering or adhesive, e.g where an item, normally machined from a casting, is made from separate items screwed or welded together.

Fettle: The act of removing irregularities from castings prior to machining or painting. Usually done with a file, sander or off-hand grinder.

Fluteless tap: A tap that produces internal threads by cold-forming, rather than by removing metal by a cutting action.

Full nut: The normal thickness nut for a given thread size. See also Lock nut.

G

Gap bed: A gap between a lathe's headstock and its bed slideways, enabling a larger diameter to be accommodated locally.

Gauge blocks: See Slip gauges

Gib strip: The thin strip of metal placed between the fixed and moving portions of a machine slide on the side with the adjusting screws in order to take the pressure from the adjusting screws and preventing them bearing on the sliding surface.

Grub screw: A small headless screw, normally with a slotted drive, typically used for fixing a bush on to a shaft. See also Set screws

H

Half nut: A nut that is split, enabling it to be opened clear of the lathe's leadscrew. When it is closed the saddle is traversed using the leadscrew. When it is open it is traversed using the rack and pinion, thus enabling the saddle to be moved more rapidly than by the leadscrew. See also Lock nut.

Honing: Similar to lapping but normally done with solid slip stones in special holders, rather than with lapping paste and split mandrels. In most cases, it is therefore only suitable for outside diameters and flat surfaces. See Lapping.

I

Idler: A gear positioned between driver and driven gears, whose purpose is either to make up the distance between the other two gears or to create a reversal. The three axles do not have to be in a straight line, and the number of teeth on the idler has no effect on the overall ratio. It can therefore be of any size that ensures a drive between the driver and the driven.

J

Jaycot tool: A tailstock mounted fixed steady used for small diameters.

Jobber drills: By far the most common form of twist drill, suitable for the majority of machining operations. Many other drills are made for special applications although, other than Stub and Long series, they are rarely found in the small workshop. See also Stub and Long series.

K

Keats angle plate: A V-angle plate for use on the lathe faceplate, typically for holding a workpiece that is too large to be accommodated in the available chucks, or for mounting it off center by an amount that is greater than can be achieved using a four-jaw chuck.

L

Lathe carrier: See Dog.

Lapping: The action of using an abrasive paste on an expanding mandrel to impart a fine finish and/or an accurate diameter, to a workpiece inside diameter. Similarly an outside diameter can be finished using an adjustable split ring. Either the workpiece or the mandrel can be rotating. See also Honing.

Live center: Originally, the lathe spindle mounted center that rotated with the workpiece and therefore did not need to be hardened. Now applied to a center that is mounted in bearings and, as a result, if mounted in the tailstock, will rotate with the workpiece. See also Dead center.

Locknut: A nut thinner than the normal thickness nut (see Full nut). Normally tightened on to the top of a full nut to assist in avoiding the fixing from working loose, of particular benefit where vibration is present. Sometimes called a 'Half nut'.

M

Mandrel: Another name for the lathe spindle. Also a spindle, chuck-mounted or between centers, on which a workpiece is placed for machining. See also Arbor.

Module: See Chapter 12 re gearing.

Morse taper: A self-holding taper used widely for mounting drill chucks and the like to machine spindles. Other tapers (Brown & Sharpe, Jarno etc) are also used but much less frequently.

N

Normalizing: Raising the temperature of a steel component to a given temperature, maintaining this for a given time, then allowing it to cool slowly. Normally used to return steel to its original state after another process, e.g the plating of a steel spring. Temperature and time are quite critical.

No volt release starter: A starter that prevents a motor from starting automatically after a power failure or other power loss, always requiring the operator to press the start button to start the motor.

O

Off-hand grinding: Grinding where the item being worked is hand-held. Wherever possible this should include additional support from the grinder table or rest.

P

Penny washer: A washer with an outside diameter that is much larger than is normal for its inside diameter.

Pinning: This occurs when small particles of a material being filed, mostly softer metals, clog the teeth of a file. This causes the particles to rub and scratch the material being worked, rather than the file removing more material.

Preload: The action of taking up any clearance present in a rolling bearing assembly and also applying a very small load while still otherwise in the unloaded state. This is done by moving one race in relation to the other, using a spring or some means of threaded adjustment. A typical application would be a machine spindle where any movement, axial or radial, would be unacceptable.

Pinion: When two gears mesh together the smaller is called the pinion, as is the gear that meshes with a rack.

Pitch circle diameter: The diameter on which a series of holes, or other items, are spaced, in most cases equally. See also Chapter 12 re gearing.

Q

Quill: The non-rotating portion of a drilling or milling machine that travels up and down with the drilling spindle.

R

Radial: Items radiating from the axis of a component, e.g. the spokes of a wheel (around the axis) or the connecting wires from the side of an electrical component (along the axis). See also Axial.

Rack: A straight bar with teeth along its edge for meshing with a gear, used for converting rotation to linear movement or vice versa.

Rake: The angle between the cutter and a line perpendicular to the face of the material being machined, most often quoted regarding the top face of lathe tools. Also applicable to drills and milling cutters. This is the face over which the material being removed passes.

Relief: The angle between the cutting edge and the workpiece that gives clearance and permits the tool to cut into the material being worked.

Roll tap: A fluteless tap that produces internal threads by cold forming rather than by removing metal by a cutting action.

S

Safe edge: The edge of a flat file that is left without teeth.

Saddle: The part of a lathe that travels along the bed of a center lathe. Also called the Carriage

Set screw: Applied to two forms of screw. Normal hexagon head screws are called set screws, as are socket set screws that are headless screws with a hexagon socket. See also Grub screws

Silver steel: A grade of steel that is relatively easy to harden. It is therefore widely used when cutting tools and the like are made in the small workshop and is readily available.

Sine bar: A precision-made bar which, together with slip gauges, is capable of setting up angles to a very high degree of accuracy.

Single cut: A file made with a single series of cutting edges creating parallel teeth

Sleeve: A tubular device for converting one diameter to another. If inner and outer diameters are made with Morse tapers, this can enable a device, such as a drill chuck with one taper, to be fitted into a machine spindle with another taper.

Slip gauges: Small rectangular blocks made in a number of thicknesses, typically 88 blocks in a metric set, and to a very high level of precision. By stacking the blocks, sizes can be created in 0.0025mm steps from 2mm up to a few hundred millimeters. Also called Gauge blocks.

Slocombe drill: Another name for a center drill.

Spinning: The action of forming thin sheet metal components, domes, bells, etc. around a former while metal and former are rotating in the lathe.

Stub drill: Drills with a flute length half that of a jobber drill. Being short they are more robust and are particularly suited to arduous tasks. They are also of benefit where space is limited, e.g. in the case of a smaller lathe.

Swing: Applied to a lathe, this is twice the center height (height of the lathe mandrel above the lathe bed) and is the maximum diameter that can be accommodated. Used in the US to describe a lathe's size.

T

Tang: The flat tongue at the small end of the taper on a drill with a Morse taper shank. The tang provides the drive to the drill rather than the grip on the taper. Also the pointed end of a file that goes into the handle.

Taper pin: A tapered pin used for positioning two parts. Being tapered it is self-holding and, unlike a dowel pin (see above) can be used in some cases for holding as well as locating. Typical use is the positioning and holding of a gearwheel on its spindle.

Tempering: Heating hardened tool steel to less than red heat, so as to relieve internal stresses. Also, and in most cases more important, makes the item less brittle and more able to withstand the stresses applied to it when in use.

Thread indicator: A dial on the lathe apron, coupled via a gear to the leadscrew. Its purpose is to establish the correct position for closing the half nut when screw cutting on the lathe.

Throat: The distance on a machine tool, such as a drilling machine, between the machining point and the part of the machine that determines the maximum distance between the machining point and the workpiece edge. In the case of a drilling machine, this is the drill spindle and machine column.

Timing belt drive: An alternative to V- or flat-drive belt systems, in which both the belt and the pulleys have teeth. This makes the pulleys resemble gears. Belt slip is eliminated and precise ratios are possible. High belt tensions are not required and power transmission is good.

Toggle clamp: A clamp using a pair of links pivoted separately at one end and having a common pivot at their other ends. Pressure is achieved by bringing the three pivot points into line.

Tommy bar: A radially mounted bar in the end of a tightening or adjusting screw, typically that included for tightening a bench vise. The bar may be captive, as in the case of a bench vise, or loose, as used for tightening a toolmaker's clamp.

Trepanning: The act of making a hole using a single point tool rotating about a central pilot pin. Usually adopted for large holes in thin sheet.

Tumbler reverse: A series of two gears mounted between the gear on the lathe mandrel and the changewheel. The assembly is mounted on a lever so that either one or two gears can be engaged, enabling a reversal between lathe mandrel and the leadscrew. In mid-position the drive is disengaged.

Turret lathe: An American term for a Capstan lathe (see above).

U

Up milling: The recommended method for the majority of milling operations (but see Down milling), where the rotation of the cutter opposes the movement of the workpiece. This avoids the workpiece being drawn under the cutter because the back lash in the feedscrew is taken up, as is possible with down milling. Sometimes known as Conventional milling.

V

Vernier: A short scale that moves adjacent to a longer scale. The shorter scale is calibrated at a slightly different pitch to that on the longer scale, which is true to size. This permits accurate adjustments or measurements, which are much finer than the divisions on the longer scale.

W

Wobbler: See Wiggler.

Wiggler: A device that enables faces or hole centers of a component to be accurately located while mounted on the machine table. This enables the position for machining to be accurately set. Mainly used on the milling machine

Woodruff key: A small half round disk that is used to provide drive between a spindle and a pulley or a gearwheel mounted on to it.

Woodruff cutter: A small milling cutter used for making a semi-circular cut into a spindle for the fitting of a woodruff key.

ABBREVIATIONS

In the past, in the engineering environment, there has been little attempt to standardize abbreviations, which means that the following abbreviations may not always be entirely accurate or consistent. The main variations are in the use of upper or lower case (e.g. alternating current as AC or ac), and the inclusion or omission of full stops (e.g. Bright Mild Steel as BMS or B.M.S.) and slashes.

In the entries below, all punctuation has been omitted and only the alphabetical characters are given.

3ph.	3-phase
A	amp
AC	alternating current
AF	across flats
ANSI	American National Standards Institute
BA	British Association (thread)
BG	Birmingham Gauge (sheet)
BGSC	Back-geared screw cutting
BMS	Bright mild steel
BSF	British Standard fine (thread)
BSI	British Standard Institute
BSP	British Standard pipe(thread)
BSW	British Standard Whitworth (thread)
BTU	British thermal unit
BWG	Birmingham Wire Gauge
CAD	computer-aided design
CAM	computer aided manufacture
cap	capacity
cb	counterbore cc cubic centimeter
cfm	cubic feet per minute
cgs	centimeter-gram-second
CI	cast iron
CLA.	center-line average
CNC	computer numerical control
CO	change over
CP	circular pitch
cps	cycles per second
crs	centers
CS	cast steel
csk	countersunk
cu	cubic
dB	decibel
DC	direct current
DE	double ended
deg	degree
dia	diameter
DIL	direct in line (PCB-mounted components)
DIN	Deutches Institut fur Normung (German Standard)
DOL	direct on line
DP	diametral pitch
DP	double pole
DPCO	double pole change over
DTCO	double throw change over
DTI	dial test indicator
dwg	drawing
emf	electromotive force
ext	external

fhp	fractional/friction horsepower
FLC	full load current
fpm	feet per minute
ft	foot
fwd	forward
GA	general arrangement drawing
gal	gallon
GM	gun metal
GRP	glass-reinforced plastic
GT	ground thread
hd	head
hex	hexagon
hp	horsepower
HSS	high-speed steel (sometimes HS)
HT	high tensile
Hz	Hertz (cycles per second)
ID	inside diameter
IEC	International Electrotechnical Commission
IEE	Institute of Electrical Engineers
imp	imperial
in	inch
ind	independent
int	internal
ISO	International Standards Organization
kg	kilograms
km	kilometer
kV	kilovolts
kW	kilowatts
l	liter
lb	pound (weight)
l.h.	left hand
LCD	liquid-crystal display
LED	light-emitting diode
lg	long
LPG	liquefied petroleum gas
mA	milliamps
Matl	material
µF	microfarad
mc	machine
MΩ	megaohm
met	metric
mH	millihenries
MIG	metal inert gas
min	minute, angle or time
mks	meter per kilogram per second
ml	milliliter
mm	millimeter
MOD	module (metric gear size)
mph.	miles per hour
mpm	meters per minute
MS	mild steel
MSG	Manufacturer's Standard Gauge (US)
MT	Morse taper
mV	millivolts
NC	normally closed
nF	nanofarad
NO	normally open
no	number
NTS	not to scale
NVR	no volt release
OD	outside diameter
OVL	overload
oz	ounce
PA	pitch angle
PB	phosphor bronze
PB	push button
PCB	printed circuit board
PCD	pitch circle diameter
pF	picofarad

psi	pounds per square inch
PSU	power supply unit
PVA	polyvinyl acetate
QC	quick change
rad	radius
rd	round
RH	right hand
rpm	revolutions per minute
rev.	reverse
SAE	Society of Automotive Engineers
SC	self-centering
sec	second, angle or time
SI	Système Internationale (d'Unités)
SP	single pole
SPCO	single pole change over
SPh	single phase
sq	square
SRBF	synthetic resin-bonded fabric
SRBP	synthetic resin-bonded paper
std	standard
SWG	Standard Wire Gauge
TC	tungsten carbide
TDC	top dead center
TE	totally enclosed
TEFC	totally enclosed fan-cooled
TiN	titanium nitride
TIR	total indicator reading
TPI	teeth per inch
TPI	threads per inch
UNC	unified coarse thread
UNF	unified fine thread
UTS	ultimate tensile strength
UV	ultra violet
V	volts
VA	volt amp
VDU	visual display unit (computer screen)
vol	volume
W	watts
Whit	Whitworth thread
wt	weight
yd	yard

USING ABBREVIATIONS

Common abbreviations may be used in text matter without being defined. Uncommon abbreviations that are used repeatedly throughout the text matter should be defined when first used.

Abbreviations should only be used on drawings where space is limited and in the case of less well-known abbreviations a reference list should be included on the drawing at a position where space is available.